Functional Analysis for Physics and Engineering

An Introduction

Functional Analysis for Physics and Engineering
An Introduction

Hiroyuki Shima
University of Yamanashi, Japan

CRC Press is an imprint of the
Taylor & Francis Group, an **informa** business
A SCIENCE PUBLISHERS BOOK

CRC Press
Taylor & Francis Group
6000 Broken Sound Parkway NW, Suite 300
Boca Raton, FL 33487-2742

First issued in paperback 2020

ISBN-13: 978-1-4822-2301-9 (hbk)
ISBN-13: 978-0-367-73738-2 (pbk)

Visit the Taylor & Francis Web site at
http://www.taylorandfrancis.com

and the CRC Press Web site at
http://www.crcpress.com

Dedication

To Rena, for being such the dearest wife,
To Yuuka and Ayaka, for being everything a father could want

Preface

Functional analysis is an essential element of multiple science and engineering fields, such as information engineering and quantum physics. For example, in the field of signal processing, the use of sophisticated analysis techniques beyond a simple Fourier analysis is necessary quite often, and abstract concepts concerning vector and function spaces must be understood. Because quantum technology and related physical sciences may require an in-depth knowledge of its theoretical foundation, we cannot escape a basic comprehension of operator theory, which is a major branch of functional analysis.

However, scientists and engineers who are involved in non-mathematical fields are unlikely to revisit the fundamental aspects of functional analysis. In conventional physics and engineering educational programs, the opportunity to learn abstract concepts that participate in functional analysis is seldom sufficiently incorporated. Moreover, many existing functional analysis textbooks are tailored to a senior-level mathematics curriculum. Therefore, "beginners" who are involved in a non-mathematics course of study may find gaining an adequate perspective of functional analysis difficult using traditional textbooks, which tend to include a substantial number of definitions, theorems, and proofs. To overcome the situation, we endeavor to delve into the applicability of functional analysis, its purpose and use, and provide a thorough yet simple understanding of the field.

The primary goal of this book is to provide readers a bird's-eye view of the key concepts of functional analysis. Topics include: topology, vector spaces, tensor spaces, Lebesgue integrals, and operators, to name a few. Each chapter concisely explains the purpose of the specific topic and the benefit of understanding it. Although exact logic, without ambiguity, is a significant aspect of any well-constructed mathematical theory, this book emphasizes simply stated key terminologies and a big picture rather than presenting rigorous definitions and proofs. It is hoped this textbook will make a unique contribution to the frequently published area of functional analysis.

8 April, 2015

Hiroyuki Shima
At the foot of Mt. Fuji, Japan

Contents

Chapter 1

Prologue

1.1 What Functional Analysis tells us

1.1.1 "Function" and "Functional analysis"

A **function** is a conceptual device that defines the relationship between two variables. The relationship is often expressed in the form of an equation or a graphical curve. In his study of the motion of an object, Isaac Newton developed a theory expressing the curvature and the area enclosed by a curve as algebraic limits. In so doing, he laid the foundation for the mathematical discipline, called **analysis**, which links shapes, computation, and physical phenomena through the concept of functions.

In light of this background, how can we define **functional analysis**? Functional analysis is the study of the relationship between one function and another through generalization of the concept of linear algebra. Our attention is paid to the distinctive nature of the **spaces**[1] formed by many functions, called **function spaces**. For instance, every function can be represented in functional analysis as a vector (arrow). Our task is to consider the manner in which the magnitude of a vector is evaluated and how different vectors are compared.

Interestingly, in the study of functional analysis, there is no unique standard for the evaluation of vectors. The criteria can be changed, depending on the circumstances under consideration. Indeed, we can set up a playground (e.g., a function space) and performers (e.g., various norms and operators) of one's choosing and thus freely create various scenarios. As a result of this adaptability, functional analysis serves as a powerful tool for a variety of mathematical problems, such as

[1] The term *space* in mathematics means a collection of components (or elements). It is different from the space that is visualized normally in everyday life, having dimensions of length, width, and height. For example, the many kinds of fruits can be said to constitute a space (fruit space, so to speak). Apples and bananas are elements of that space.

Figure 1.1: A watermelon, an apple, and a banana are elements of the "fruit space".

the solution of partial differential equations. This is the reason why even students of non-mathematics majors feel motivated to learn the subject.

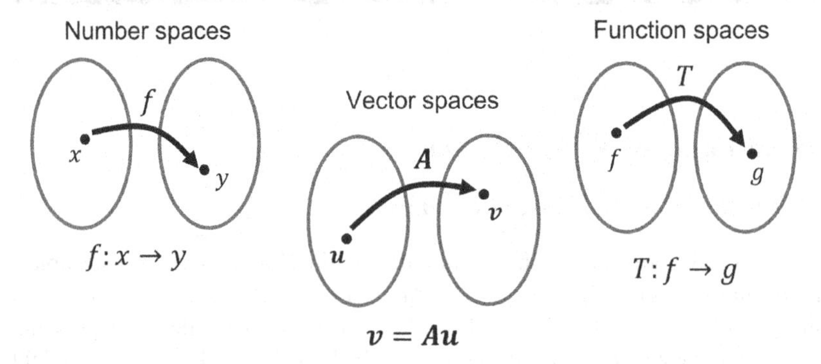

Figure 1.2: A function $y = f(x)$ relates a number x to the other number y. A matrix $\boldsymbol{A}$ relates a vector $\boldsymbol{u}$ to $\boldsymbol{v}$. In a similar manner, an operator T relates a function f to the other function g.

Keypoint: Functional analysis is the study of the relation between functions and/or vectors.

1.1.2 Infinite-dimensional spaces

The most important factor in learning functional analysis is to understand the nature of **infinite-dimensional space**. Speaking intuitively, a function can have an infinite number of **degrees of freedom**, because it can take any shape, as a result of its flexibility.[2] This number of degrees of freedom corresponds to the **dimension** of the

[2]For example, when a function is expressed as a graphic curve, it can be bent or stretched in any way. When it is expressed as an algebraic equation, it can be changed by varying the values of each term's coefficients.

space that is the stage of the function. Hence, there is a need for a theory that treats a function as a vector of infinite-dimensional space. However, the world of infinite dimensions is by no means simple and is not obtained by merely extrapolating a finite number of dimensions to infinity. There is, more often than not, profound disagreements in the nature of functions (or vectors) between those studied in elementary linear algebra (= finite dimension) and in functional analysis (= infinite dimension). Therefore, to examine the relationship between functions in infinitely many dimensions, it is important to consider the functions carefully using abstract concepts.

> **Keypoint:** Functional analysis helps us to understand the nature of infinite-dimensional spaces.

1.1.3 Relevance to quantum physics

Another remarkable feature of functional analysis is its close relationship with physics. In particular, **quantum mechanics** has had an especially close association with functional analysis, and the two have evolved together through mutual stimulation in historical perspective. For example, the **canonical commutation relation**, a basic principle of quantum mechanics, is expressed as follows,[3] using square matrices $\boldsymbol{P}$ and $\boldsymbol{Q}$.

$$\boldsymbol{PQ}-\boldsymbol{QP}=-i\boldsymbol{I}. \tag{1.1}$$

The point to take note of here is that the matrices $\boldsymbol{P}$ and $\boldsymbol{Q}$ satisfying this relationship must be of infinite dimension.[4] If that is the case, then it is possible to construct a matrix satisfying the canonical commutation relations by introducing the matrix $\boldsymbol{A}$ given by

$$\boldsymbol{A}=\begin{pmatrix} 0 & 0 & 0 & 0 & \cdots \\ \sqrt{1} & 0 & 0 & 0 & \cdots \\ 0 & \sqrt{2} & 0 & 0 & \cdots \\ 0 & 0 & \sqrt{3} & 0 & \cdots \\ \vdots & \vdots & \vdots & & \ddots \end{pmatrix} \tag{1.2}$$

and taking $\boldsymbol{P}$ and $\boldsymbol{Q}$ to be, respectively,[5]

$$\boldsymbol{P}=-i\frac{(\boldsymbol{A}^T-\boldsymbol{A})}{\sqrt{2}} \tag{1.3}$$

---

[3] $\boldsymbol{I}$ is the **identity matrix**, which is also called **unit matrix**.

[4] By looking at the sum of the diagonal elements on both sides, it is easily seen that the expression (1.1) cannot be satisfied if $\boldsymbol{P}$ and $\boldsymbol{Q}$ are finite matrices of order n. The diagonal sum on the right-hand side is $-in$, while, on the left hand side, the diagonal sum of $\boldsymbol{PQ}-\boldsymbol{QP}$ is always zero, since the sum of the diagonal elements of $\boldsymbol{PQ}$ and $\boldsymbol{QP}$ are equal.

[5] The upper right subscript T of $\boldsymbol{A}^T$ means that it is the **transpose matrix** of $\boldsymbol{A}$.

and

$$Q = \frac{A^T + A}{\sqrt{2}} \tag{1.4}$$

In other words, the introduction of infinite-dimensional vectors and matrices is indispensable for formulating quantum theory. But their mathematical properties cannot be understood merely through simple analogies from finite-dimensional cases. Finite- and infinite-dimensional spaces are different in essence. This is another factor motivating non-mathematics majors to study functional analysis.

Keypoint: Functional analysis gives a theoretical basis of quantum mechanics.

The foregoing description provides a bird's-eye view on the reason why functional analysis is an important branch in mathematics. In the remaining sections of this chapter, we explain the merits and knowledge which you will obtain by studying functional analysis from the three different perspectives: limits, infinite dimensions, and relationship with physics.

1.2 From perspective of the limit

1.2.1 Limit of a function

Again, what is functional analysis? Before attempting to answer this question, let us remove the adjective "functional" and examine the essence of "**analysis**".

Analysis is one of the three major branches of mathematics, along with **algebra** and **geometry**, centering on the understanding of the concept of limits. Isaac Newton, the founder of mathematical analysis, considered the motion of an object in the limiting condition, when the time is zero, and elucidated laws of physics that govern nature. According to his findings, the motion of an object with a mass at any moment is expressed by the equation:

$$\text{Mass} \times \text{Acceleration} = \text{Force}. \tag{1.5}$$

The acceleration mentioned here is the instantaneous rate of change of velocity, and velocity is the instantaneous rate of change of displacement. This instantaneous rate of change is an example of what is known today as the concept of limit.

Instantaneous rate of change is defined as the limit of the ratio $\Delta x/\Delta t$ when Δt becomes arbitrarily close to zero, assuming Δx to be the displacement in the finite time interval Δt. However, the actual value of this limit cannot be measured directly through experiments, because a finite time interval is necessary in actual measurement situations and shrinking this interval to zero would be equivalent to not measuring at all. Consequently, to estimate the instantaneous rate of change, an array of measurements:

$$\{\Delta x_1, \Delta x_2, \ldots\} \tag{1.6}$$

Figure 1.3: Imagine the conclusion from Newton's equation of motion (1.5). A force exerted on a small mass (= a golf ball) produces large acceleration, while the same force acting on a large mass (= a truckload) results in only slight acceleration.

taken over a number of finite time intervals:

$$\{\Delta t_1, \Delta t_2, \cdots\} \tag{1.7}$$

is created, and the behavior of the ratio $\Delta \boldsymbol{x}_n/\Delta t_n$, when the time interval approaches zero, is examined. Defining this limiting operation, known as approaching zero, unambiguously is the principal task of mathematical analysis.

Calculation of limits is taught as a matter of course in schools today. However, it is the wisdom of many great minds that has enabled the human intellect to handle limits without contradiction. In fact, the calculations that we can execute run at most a finite number of times.[6] Therefore, what kind of result will be obtained by repeating a predetermined procedure an infinite number of times? Is there a fixed goal? If so, then where is it? Without mathematical analysis, these questions can never be answered.

> **Keypoint:** Mathematical analysis is a study that deals with finding the limit of a given function.

[6]This is true even with the use of large-scale state-of-the-art computers.

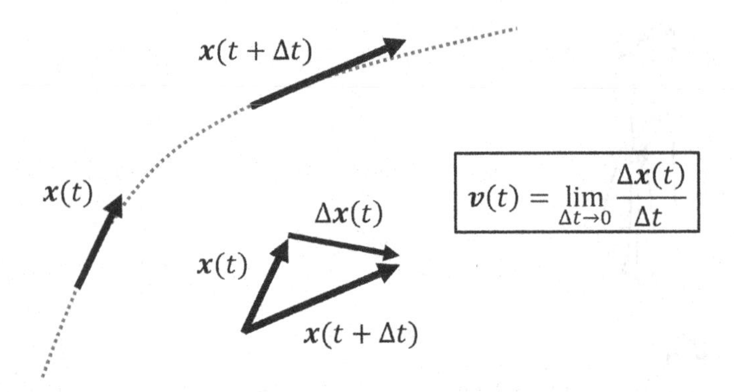

Figure 1.4: Change in the time-dependent vector $\boldsymbol{x}(t)$ during the time interval Δt. The velocity vector $\boldsymbol{v}(t)$ is defined by the limit described.

1.2.2 What do you want to know: the limit's value or existence?

The paradox of Achilles and the tortoise discussed in ancient Greece is famous as an example of the difficulty of dealing with limits. This paradox explores whether the fleet-footed Achilles can catch up with a turtle that is walking ahead of him.

Let x_1 be the location of the turtle at a specific time t_1. Let us assume that Achilles, who is pursuing the turtle from behind, arrived at this point x_1 at time t_2. At this time t_2, the turtle would have advanced to the point x_2 a little ahead. Subsequently, at the time t_3 when Achilles arrives at x_2, the turtle would have moved to x_3 further ahead. If this operation is continued repeatedly, the turtle will always be a step ahead of Achilles, and it appears that Achilles can never catch up with the turtle. However, in the real world, even an ordinary person who is much slower than Achilles can overtake the turtle easily. Hence, we ask what is the error in the foregoing discussion?

The key to this paradox lies in the fact that although the number of terms of the sequence $t_1, t_2, \cdots$ is infinite, $\lim_{n\to\infty} t_n$ exists as a finite limit. However large the number of terms for which the value of t_n is determined, that value can never exceed a specific value, even though the specific value cannot be established concretely. It is therefore incorrect to conclude that such a specific value does not exist. To put it differently, it is not true that Achilles will never keep up with the turtle, no matter how long he runs.

Rather, the truth is that as long as we rely on the calculation of successive values of t_n to determine the point where Achilles overtakes the turtle, such calculations will reach an inevitable deadlock situation, namely, that the overtaking point can never be found. Thus, mathematical analysis is a very useful aid in understanding the concept of limits, which might be difficult to grasp intuitively.

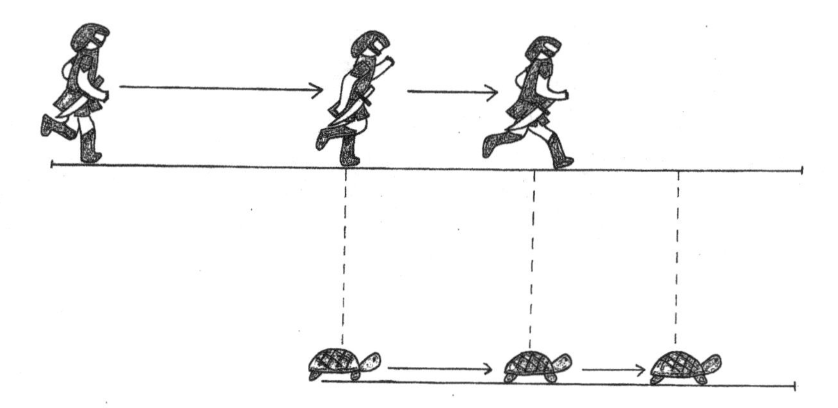

Figure 1.5: Diagram for the paradox of Achilles and a tortoise. Can the fast runner Achilles catch up a tortoise?

> **Keypoint:** It is often impossible to specify "*the value*" of the limit of an infinite sequence. Using the analysis, however, we can assure "*the existence*" of the limit at least.

1.2.3 Indirect approach to the limit

In mathematical analysis, a rigorous treatment of limits is performed using the **epsilon-delta (ε-δ) logic**. For example, suppose that a real function $f(x)$ is continuous at a particular real number $x = a$. This fact is written using ε-δ logic as follows[7]:

$$\forall \varepsilon > 0, \exists \delta > 0 : |x-a| < \delta \Rightarrow |f(x) - f(a)| < \varepsilon \tag{1.8}$$

Without using symbols, this expression reads as follows:

"For every positive number ε, there exists a positive number δ that fulfills the following condition: *If* $|x-a| < \delta$, *then* $|f(x) - f(a)| < \varepsilon$."

However, this statement is very different from the language of everyday use and might not be comprehensible in a single reading.

The reason the ε-δ logic is difficult to understand could be that the roles of the two people in the argument (the person assigning ε values and the one assigning δ values) are confused. In the ε-δ logic, the speakers corner each other, so to speak, as in the game of rock-paper-scissors. One of the partners first suggests a value for ε as small as he or she likes. Upon hearing the choice of ε, the other partner gives a value

[7] Here the symbols $\forall$ and $\exists$ mean "for any" and "there exists", respectively.

Figure 1.6: The ε-δ logic is analogous to cheating in the game of rock-paper-scissors. The δ-player always wins.

for δ for which x values satisfying $|x-a| < \delta$ yield values of $|f(x) - f(a)|$ that are smaller than ε. Mortified, the first partner proposes an even smaller value for ε and the argument proceeds on these lines. However, small a value of ε is proposed, the other partner can come up with a value for δ for which x values satisfying $|x-a| < \delta$ yield smaller values of $|f(x) - f(a)|$, making it appear a rather malicious game. Repeating this unproductive argument countless times finally creates the possibility of ascertaining the continuity of the function $f(x)$ at $x = a$.

> **Keypoint:** The ε-δ logic is a cheating in the "rock-paper-scissors" game; the δ-player always makes his/her move later than the ε-player.

This ε-δ logic is a very roundabout line of argument. It avoids tackling the target value of $x = a$ and continues to step around the value instead. The logic behind this reasoning is that in order to know the continuity of $f(x)$ at $x = a$, it is sufficient to know the behavior of the function in the immediate vicinity of $x = a$. This manner of not shooting the arrow directly at the target, but gradually closing in from the periphery is an essential feature of mathematical analysis.

Using the ε-δ logic, it is possible to deduce the continuity of $f(x)$ at $x = a$, even if the value $f(a)$ is not reached.

1.2.4 *Function as the limit of mathematical operations*

Having established functions in the usual sense as the basis of mathematical analysis, we now look at them from the perspective of the limit of a mathematical operation.

Originally, a function was a conceptual device to generalize the individual relationship of one variable with another. For example, suppose that an electric current of 150 mA flows through a device with a voltage of 1 V applied to it. Suppose that similarly the measurement data of 300 mA of current with voltage of 2 V, 450 mA of current with voltage 3V, ... are obtained. We generalize this empirical fact (knowledge from a finite number of attempts) and infer that the relationship $y = 0.15x$ holds between voltage x [V] and current y [A] (Ohm's law). That is, the relational expression that holds for finite values of x and y is considered to be established even for the continuous variables x and y, thus making the law applicable to the infinite-dimensional case.

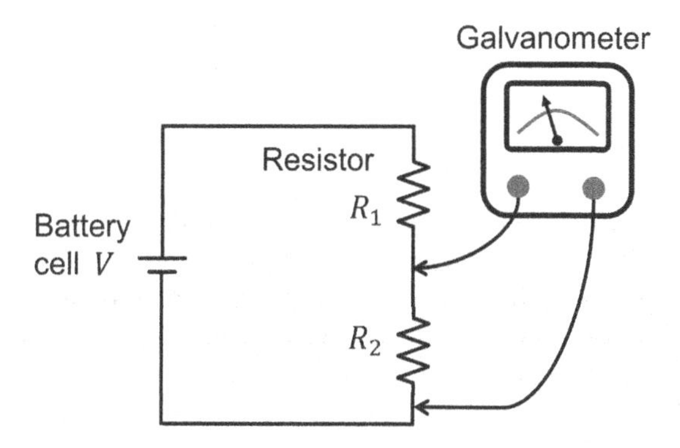

Figure 1.7: Electric circuit.

This example is of a function that is induced from experience. However, functions deduced from assumptions are more important in mathematics. For example, suppose that as a result of theoretical analysis, the movement of an object is expressed as $x = F(t)$ where x is the position of the object, and t is the time. Knowing the nature of $F(t)$ would mean understanding the movement of the object. What does knowing the function mean? In general, the identity of any function cannot be known unless infinitely many mathematical procedures are followed.

Let us consider some well-known elementary functions. Trigonometric functions such as $\cos(\omega t)$, the **Bessel function**[8] $J_\nu(x)$, and the **Jacobi elliptic function**[9] $\mathrm{sn}(u,k)$ are useful functions that have been assigned special symbols. Therefore, their specific values and fundamental properties (calculus formulae, sum formula, recurrence formula, and the like) have already been examined exhaustively. How can such fundamental properties be investigated? For this, it is necessary to give a specific identity to the target function, such as expressing it in the form of a power series expansion or an indefinite integral or by specifying the differential equation that governs the function. For example, although the exponential function $f(x) = e^x$ is defined as a primitive function that satisfies the equation

$$f(x+y) = f(x)f(y), \tag{1.9}$$

its various characteristics are not made clear from this definition alone. Hence, procedures such as expressing this function as the sum of an infinite series

$$f(x) = \lim_{N\to\infty}\left(\sum_{n=0}^{N}\frac{x^n}{n!}\right), \tag{1.10}$$

or regarding it as the solution to the differential equation

$$\frac{df}{dx} = f(x), \tag{1.11}$$

when assigning the initial value $f(0) = 1$ are adopted. The latter method, in particular, the method that focuses on the differential equation that governs a function as a means to exploring its identity, is in line with the idea of classical mechanics that emphasizes the relationship between the instantaneous movement of an object (speed and acceleration) and the locus traced by it.

The point to be noted here is that differential equations such as expression (1.11) (referred to in classical mechanics as equations of motion) are balancing relationships established with respect to an infinitesimal x. If this balancing relationship is approximated with finite values of x (i.e., if the differential is approximated by the difference), the approximate solution of the function (motion) will be obtained through finite computations. This limit is known as the true function (motion). In order to know the specific form of a function, whether in calculating the limit of an infinite series or in calculating the infinitesimal limit of a differential, it would be necessary to carry out infinitely many rounds of calculation except in particularly simple cases (such as finite order polynomials). Generally, the true form of a function (motion) can only be perceived as the limit of its approximations.

---

[8]The **Bessel function** (of the first kind) $J_\nu(x)$ is defined by the solutions to the **Bessel differential equation** $x^2y'' + xy' + (x^2 - n^2)y = 0$ with integers n. It can also be defined by the integral $J_m(x) = (1/\pi)\int_0^\pi \cos(x\sin\theta - m\theta)d\theta$.

[9]**Jacobi elliptic function** is defined by the inversion of the integral $F(\phi,k) = \int_0^\phi \left(1 - k^2\sin^2 t\right)^{-1/2} dt$ where $0 < k^2 < 1$; namely, $\mathrm{sn}(u,k) \equiv \sin\phi$ where $\phi = F^{-1}(u,k)$.

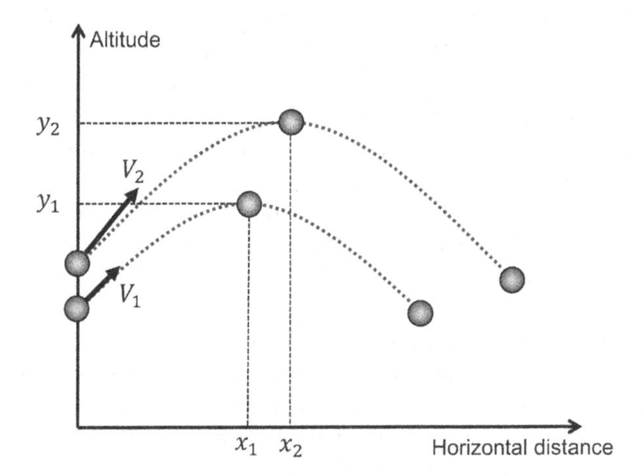

Figure 1.8: Projectile pathway.

In order to be able to regard a function as a limit, it is necessary to develop the concept of the limit of a sequence, which is learned at the beginning of mathematical analysis, and to form a theory that will help in visualizing the limit of a sequence of functions. This means rising one level higher from:

The theory for calculating the limit of "*an individual function*"

to:

The theory for the limit of "*an infinite sequence of functions*".

The higher level above-mentioned is the field of functional analysis.

1.3 From perspective of infinite dimension

1.3.1 Topology of a space

In general, the quantitative laws of science are expressed as relationships between variables. As we go deeper into the theory of analysis, the values of variables change from real numbers to complex numbers, then to vectors and polynomials, and finally to functions, which are themselves regarded as possible values of variables. Functional analysis endeavors to develop mathematical analysis in this wonderful space, function space, where functions are considered as variables.

The elements in a function space have infinitely many **degrees of freedom**. To discuss limits or convergence, it is necessary to have rules that determine the proximity or remoteness of the elements (the distance between elements, so to speak). The standard that determines the difference between the elements is called the **topology** of the space. As will be learned later in this book (see Chapter 2), it is possible to set up various kinds of topologies in an infinite-dimensional function space. Depending on the setting of the topology, the character of the space changes dramatically. Our interest will therefore be to determine the topology that will yield a meaningful space. Establishing mathematical analysis in a function space begins with defining the relationships among the elements.

When discussing the topology (structure) of an abstract space, it is often convenient to compare the elements of the space with vectors in our familiar three-dimensional **Euclidean space**. It is the aim of functional analysis to carry out such conceptual operations in the abstract infinite-dimensional space called the **function space**. It will then be possible to introduce the geometric concepts of Euclidean space, such as the length or inner product of vectors, into the function space. However, simple analogies are not always helpful, because a function space is an infinite-dimensional space, into which an in-depth study of limits and convergence is essential for the introduction of geometric concepts.

1.3.2 Length of a vector

The essential difference between Euclidean space and function space is that the latter is infinite-dimensional. This infinite-dimensional nature is a natural requirement, because the elements of the space are functions. Since functions undergo transformations easily, there are countless varieties of them. Therefore, in order to represent a function as a vector starting at the origin and ending at a particular point in the space, the vector must have infinitely many degrees of freedom. Here, the term *degrees of freedom* refers to the number of mutually independent variables that determine a vector. This number is the number of axes spanning the vector space, in other words, the dimension of the space. Thus, the function space must be infinite-dimensional.

To visualize infinite-dimensional vector space, we will review the Euclidean space of three dimensions. Here, a point in space is represented by its coordinates, written as

$$(x_1, x_2, x_3). \tag{1.12}$$

The fact that there are three coordinate components is directly linked to the number of dimensions of the space (the number three in this case). Therefore, an infinite-dimensional space can be expressed, if the number of coordinates is infinite. In other words, a vector such as

$$\boldsymbol{x} = (x_1, x_2, x_3, \cdots x_n, \cdots) \tag{1.13}$$

consisting of a countless number of x_i's is regarded a point, and a set of such points is considered a space.

Although it might seem that the matter ends there, the infinite-dimensional space created in this manner is not very useful, because the length of a vector (or the distance between two points) is not well defined. In the case of three-dimensional

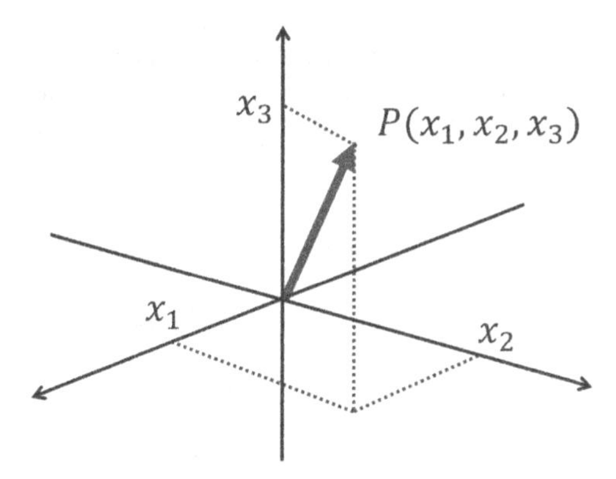

Figure 1.9: A point P in the three-dimensional Euclidean space is represented by three variables: x_1, x_2, x_3.

space, the length of a vector $\boldsymbol{x} = (x_1, x_2, x_3)$ can be expressed as

$$\sqrt{|x_1|^2 + |x_2|^2 + |x_3|^2}. \tag{1.14}$$

If this expression were to be used as it is, the length of a vector in an infinite-dimensional space could be defined as

$$\sqrt{|x_1|^2 + |x_2|^2 + |x_3|^2 + \cdots + |x_n|^2 + \cdots}, \tag{1.15}$$

but this infinite sum is not necessarily convergent. If the sum is divergent, the distance between the two points $\boldsymbol{x}$ and $\boldsymbol{y}$ in the space (i.e., the length of the vector $\boldsymbol{x} - \boldsymbol{y}$) cannot be discussed, and hence the concept of limits cannot be used. Thus, the tool of mathematical analysis is not very helpful.

So what can be done? It is possible to try to redefine the quantity that corresponds to the length of the vector, but such a strategy is not being adopted here. In the infinite summation expression of (1.15), it would be unreasonable to gather together all the countless number of x_i's. We therefore consider collecting only those

$$\boldsymbol{x} = (x_1, x_2, x_3, \cdots, x_n, \cdots) \tag{1.16}$$

that will make the summation convergent, and the quantity corresponding to the length of $\boldsymbol{x}$ is represented by the symbol

$$\|\boldsymbol{x}\| = \sqrt{|x_1|^2 + |x_2|^2 + |x_3|^2 + \cdots + |x_n|^2 + \cdots}. \tag{1.17}$$

The space that contains only those $\boldsymbol{x}$'s that are suitable is known as the ℓ^2 **space** in functional analysis, and the quantity corresponding to the length, as defined by the expression (1.17), is called the ℓ^2 **norm**.[10] The length can also be defined in other manners, and with every change in the definition of the length, a different type of infinite-dimensional space with a different topology is created. The ℓ^2 space described above is an example of the most basic among those spaces.

> **Keypoint:** Our intuitive understanding of *finite* dimensional spaces do not apply on their own to *infinite* dimensional spaces.

1.3.3 Size of a function

Similar arguments will be made regarding the set of functions. Let us consider the entire set of continuous functions $f(x)$ of the real variable x $(0 \le x \le 1)$. This function space is written using the symbol C, as in $C[0,1]$, for example.[11] Now we regard the elements of the $C[0,1]$ space as a vector. While various options can be considered as expressions for the length of the vector [i.e., the magnitude of the function $f(x)$], the following two are of particular importance. The first is to define the magnitude of $f(x)$ as

$$\|f\| = \max_{0 \le x \le 1} |f(x)| . \tag{1.18}$$

This expression is easy to understand intuitively and may give the feeling that it does indeed measure the magnitude of the function. A second method is to define the length as

$$\|f\| = \sqrt{\int_0^1 |f(x)|^2 dx}. \tag{1.19}$$

Since the latter expression does not seem a natural way to express the length, it is explained below.

Suppose the function $f(x)$ is defined over the set of natural numbers $\{1,2,\cdots,n\}$, rather than over the set of real numbers. In this case, the sequence $\{f(1), f(2), \cdots, f(n)\}$ can be regarded as a vector component of the n-dimensional space. Hence, it would be natural to assume its length to be

$$\sqrt{|f(1)|^2 + |f(2)|^2 + \cdots + |f(n)|^2} = \sqrt{\sum_{i=1}^{n} |f(i)|^2}. \tag{1.20}$$

Let us extend the domain of x from the set of natural numbers $\{1, 2, \cdots, n\}$ to the interval $[0, 1]$ of real numbers. Since integration is a natural extension of summation, the expression (1.19) is a natural way to represent the magnitude of the function. The

---

[10] The superscript "2" in the symbol "ℓ^2" indicates that the sum of $|x_i|^2$ with $(i = 1, 2, \cdots)$ is being taken in the expression (1.17) that defines the norm $\|\boldsymbol{x}\|$.

[11] The symbol C is an abbreviation for "continuous", and the symbol $[0,1]$ means that the domain of x is the closed interval $0 \le x \le 1$.

quantity given by the expression (1.19) is called the L^2 **norm**, and the space created by gathering together only those functions $f(x)$ that would make the L^2 norm take a finite value is called the L^2 **space**. while the ℓ^2 norm discussed earlier represents the size of an *infinite number sequence*, the L^2 norm defined here represents the size of a *function*.

> **Keypoint:** Many different definitions are allowed for the length of elements in a space (i.e., the magnitude of functions in a functional space).

Two of the most well-known classes of infinite-dimensional spaces, the ℓ^2 spaces and L^2 spaces, have been introduced above. It must be noted that the former space is a set of sequences, and the latter is a set of functions. As will be discussed later, these two spaces are, in fact, equivalent mathematically. They are both the same infinite-dimensional space described using different expressions.[12] Furthermore, it is this space (called **Hilbert space**) that serves as the platform for formulating quantum theory.

1.3.4 Small infinity and large infinity

We mentioned earlier that the elements of the function space $f(x)$ have infinitely many degrees of freedom. It is thus pedagogical to mention the fact that there exists a distinction between small and big infinities. The smallest infinity is the infinity that can be counted using natural numbers. The infinite set having the same **cardinality** (= size) as the set of all natural numbers (represented by the symbol "$\mathbb{N}$") is called a **countable set**, and its cardinality is denoted by[13] $\aleph_0$. The set $\mathbb{Z}$ of all integers and the set $\mathbb{Q}$ of rational numbers are also countable sets, just as $\mathbb{N}$ is. However, the set $\mathbb{R}$ of real numbers has a higher cardinality than $\aleph_0$ (as proved by **Cantor's theorem**) and its cardinality is written as $2^{\aleph_0}$ (or designated by c). The set $\mathbb{C}$ of complex numbers is equivalent to $\mathbb{R} \times \mathbb{R}$, but its cardinality is equal to $2^{\aleph_0}$.

How big is the infinite number of degrees of freedom of the function $f(x)$? You may think of at first glance that it has cardinality of $2^{\aleph_0}$, because a value $f(x)$ is determined once you specify the real number x in the set $\mathbb{R}$ that has cardinality of $2^{\aleph_0}$. However, this estimation is incorrect. Surprisingly, the number of degrees of freedom of the functions that we usually deal with is the countably infinite $\aleph_0$, similarly to $\mathbb{N}$.[14]

---

[12] In mathematics, the term **expression** means a written concrete mathematical model for an abstract system defined by axioms. For example, the coordinate representation of a vector, by introducing coordinates into the axiomatically defined vector space, is an expression of a vector. Speaking plainly, it expresses abstract and hard-to-understand concepts in concrete and easy-to-grasp fashion.

[13] $\aleph$ is a Hebrew letter, read as *Aleph*.

[14] This fact can be deduced in the following manner. Suppose $f(x)$ is a continuous function and the domain of the variable x is the open interval Ω in the set $\mathbb{R}$. Then, for any real number $x \in \Omega$, it is possible to

This discussion gives us a caution that, in order to treat infinite-dimensional space without contradiction, we need to create a robust theoretical system beyond simplistic analogy from finitely many dimensions. In other words, fundamental concepts such as space or vector, that have been grasped intuitively until now, must be rigorously defined with respect to infinitely many dimensions. Furthermore, the similarities and differences between the theories of finite and infinite numbers of dimensions need to be made clear. This is the focus of functional analysis.

Keypoint: Caution ... The nature of infinity goes far beyond our intuition.

1.4 From perspective of quantum mechanical theory

1.4.1 Physical realization of the operator theory

Reviewing the abstractions described thus far, the theory of functional analysis could appear to be highly axiomatic in nature. However, it must not be forgotten that the theories of natural science were not born from mere speculation, but from necessity.

Figure 1.10: Schrödiner's cat vs. Newton's apple.

In the present case, it is quantum theory that imbued life into functional analysis. Historically as well, quantum theory that governs the subatomic world, would be incomplete without operator analysis, which forms the basis of functional analysis; See Section 10.1.5. In elementary mathematical analysis, the function itself is the

find a sequence $\{x_m\} \subset \mathbb{Q}$ of rational numbers very close to x such that $\lim_{m\to\infty} x_m = x$. Further, since $f(x)$ is continuous, we have $f(x) = \lim_{m\to\infty} f(x_m)$. Hence, to determine $f(x)$ uniquely, it is sufficient to specify in advance, the value of $f(x_m)$ for every set $\mathbb{Q} = \{x_m\}$ of rational numbers belonging to Ω. It is not necessary to specify $f(x)$ for all real numbers x. Thus, it is seen that the number of degrees of freedom of a continuous function (the number of variables whose values can be determined freely) is only countably infinite.

device that operates on a variable. At a higher level, we now have to think about the device that will operate the function. The latter device is called an **operator** in functional analysis. This concept of operator is needed for the development of quantum theory, because it is the operator that serves as the physical quantity in quantum theory.

In classical mechanics, physical quantities are represented by real numbers. For instance, if the state of an object can be expressed using n physical quantities, the state space is an n-dimensional real vector space $\mathbb{R}^n$, and the change of state of the object (motion) will be expressed by the locus of the moving particles (curve) in $\mathbb{R}^n$. Correspondingly, in quantum theory, the operator is considered to be the physical quantity. The reason for this practice is the discovery that some physical quantities showed strange behaviors in experiments, which had been unknown in the era of classical theory. For instance, the energy of molecules and electrons often take only discrete values, although their classical counterparts take continuous values. More curiously, there are situations in which it is not possible to establish theoretically the value of a particular physical quantity. Here, the problem is not with the accuracy of measurements. No matter how precisely measurements are made, some situations arise where it is impossible to determine a value theoretically.

To explain these strange experimental facts, the premise that physical quantities can be determined as continuous real values is discarded in quantum theory. This means that the state space is also not a real vector space spanned by axes corresponding to each physical quantity. Consequently, the geometric conception of nature that Galileo had visualized in the past undergoes a fundamental change. Namely, we have to abandon the traditional idea that the presence and motion of objects are always equivalent to the locus of the physical quantities determined by measurements.

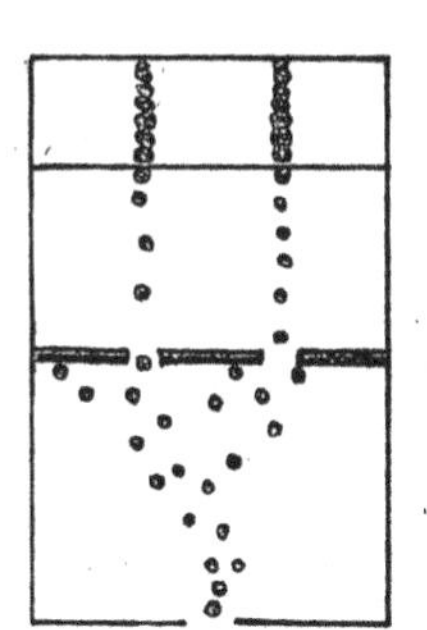
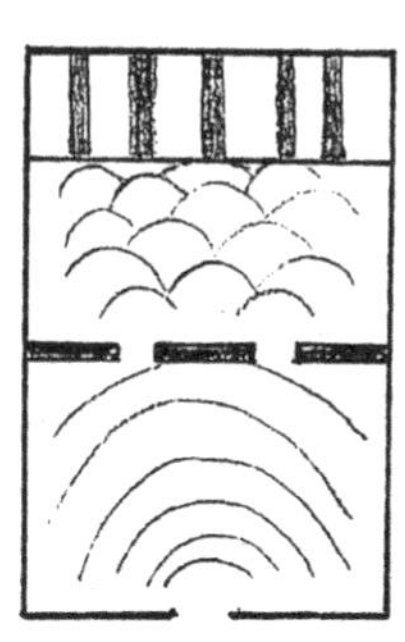

Figure 1.11: Left: Motion of classical particles. Those passing through a screen with two slits will land directly behind the slit. Right: Motion of quantum waves. The waves pass through both slits at once and interfere with each other, making a periodic interference pattern.

In quantum theory, we should consider the following two facts separately: the presence/motion of a substance and the process of measuring physical quantities through observations. First, the state of the substance is considered to be an infinite-dimensional vector. This vector moves, operated on by an operator. It is the goal of this theory to identity the operator as a physical quantity. Thus, the physical quantity is an operator. Since the operator itself is not a numerical value, it is necessary to define the mapping from the combination of the arrow and the operator to a number to know the value of the physical quantity. This is the observation of physical quantity. In quantum theory, the value of the physical quantity appears only after the state of the substance has been observed.

1.4.2 Difficulty in choosing a basis in infinite dimension

The operators generally dealt with in quantum theory are **non-commutative**. That is, two operators A_1 and A_2 denoting two different physical quantities do not satisfy the equation $A_1A_2 = A_2A_1$. The most basic theory for expressing non-commutative algebra is the theory of matrices, or, **linear algebra**. Functional analysis can thus be regarded as an extension of finite-dimensional linear algebra to infinitely many dimensions. However, not all of the techniques and concepts used in finitely many dimensions can be extended to infinitely many dimensions. The inherent difficulties with infinitely many dimensions are mentioned below.

The central theme of linear algebra is to select an appropriate base and represent a linear mapping by means of a simple matrix. The ultimate aim is to reduce a linear mapping to a proportionality rule. Given below is a linear mapping of A from the complex linear space $\mathbb{C}^n$ to $\mathbb{C}^n$ itself. The **eigenvalue problem** of A is to determine the special $\boldsymbol{\lambda}$ and $\boldsymbol{u}$ that satisfy the equation

$$A\boldsymbol{u} = \boldsymbol{\lambda}\boldsymbol{u}, \tag{1.21}$$

and translate the linear mapping A into a proportionality rule (multiplication by the complex number $\boldsymbol{\lambda}$). If this has n independent solutions, i.e., the **eigenvalues** λ_j and its **eigenvectors** $\boldsymbol{e}_j$ $(j = 1, 2, \cdots, n)$, then, by choosing the base $\{\boldsymbol{e}_1, \boldsymbol{e}_2, \cdots, \boldsymbol{e}_n\}$, A can be reduced to a set of n independent proportionality rules. In fact, if the vector $\boldsymbol{u}$ is expanded as

$$\boldsymbol{u} = \sum_{j=1}^{n} u^j \boldsymbol{e}_j, \tag{1.22}$$

and multiplied by A, then it can be written as

$$A\boldsymbol{u} = \sum_{j=1}^{n} u^j A\boldsymbol{e}_j = \sum_{j=1}^{n} \lambda_j u^j \boldsymbol{e}_j. \tag{1.23}$$

When expressed as a matrix, this becomes

$$A\begin{pmatrix} u^1 \\ \vdots \\ u^n \end{pmatrix} = \begin{pmatrix} \lambda_1 & \cdots & 0 \\ \vdots & \ddots & \vdots \\ 0 & \cdots & \lambda_n \end{pmatrix}\begin{pmatrix} u^1 \\ \vdots \\ u^n \end{pmatrix}. \tag{1.24}$$

In the case of finite-dimensional matrices, the kind of linear mappings that can be reduced to the proportionality rule has already been well studied. However, there are limitations to generalization when it comes to infinite-dimensional space. Since the number of dimensions is not finite, it is particularly difficult to judge the excess or deficiency of the number of eigenfunctions through numerical calculations. If the number n of dimensions is finite, then finding n independent eigenfunctions is sufficient. However, in the case of infinitely many dimensions, even if infinitely many eigenfunctions are obtained, it is not possible to determine whether they are sufficient by comparison with the number of dimensions. In the eigenvalue problem, there may be cases where eigenfunctions cannot be found at all and others where excessively many are found.

A general theory exists only in the case of **self-adjoint operators**. The self-adjoint operator $A^*$of A defined by

$$(\boldsymbol{u}, A^*\boldsymbol{v}) = (A\boldsymbol{u}, \boldsymbol{v}) \tag{1.25}$$

is equal to A itself.[15] The notation $(\boldsymbol{a}, \boldsymbol{b})$ here signifies the **inner product** of the two vectors $\boldsymbol{a}$ and $\boldsymbol{b}$. However, also in the theory of self-adjoint operators, a simplistic analogy with finitely many dimensions often fails.

1.4.3 Plane waves are anomalous wave functions

As a typical example of a finite-dimensional analogy that does not work, we consider the **plane wave**, the "eigenfunction" of the momentum operator. The reason for using quotation marks is the strange properties of the plane wave in infinite-dimensional space, as described below.

Suppose the value of momentum for the plane wave $\exp[i(\boldsymbol{k}\cdot\boldsymbol{x} - \omega t)]$ is estimated by operating $p_j = i\hbar\partial/\partial x^j$ on it. When integrated according to the definition, this becomes the divergent integral

$$\int_{\mathbb{R}^3}\left[-i\hbar\frac{\partial}{\partial x^j}e^{i(\boldsymbol{k}\cdot\boldsymbol{x}-\omega t)}\right]\cdot e^{-i(\boldsymbol{k}\cdot\boldsymbol{x}-\omega t)}d\boldsymbol{x} = \int_{\mathbb{R}^3}\hbar k_j d\boldsymbol{x} \;\rightarrow\; \infty. \tag{1.26}$$

Although the plane wave is an eigenfunction of the momentum operator, why is the integral indicating its observed value found to be divergent? The answer is that the plane wave is not a wave function in the correct sense. That is to say, when the plane wave is represented by a vector in the infinite-dimensional space, the norm of the vector becomes divergent.

In quantum theory, the presence of a substance is represented by an infinite-dimensional vector. Since this vector must exist as an entity in infinite-dimen -sional space, it must have a finite length. In the framework of quantum theory, when a function $f(x)$ is represented as a vector in infinite-dimensional space, the magnitude (norm) of the vector is generally defined as

$$\|f\| = \left(\int_{\mathbb{R}^3}|f(\boldsymbol{x})|^2 d\boldsymbol{x}\right)^{1/2}. \tag{1.27}$$

[15]The exact definition of a self-adjoint operator will be given in Section 10.3.3.

Accordingly, when the size of the vector corresponding to the plane wave is determined, it becomes

$$\left\| e^{i\boldsymbol{k}\cdot\boldsymbol{x}} \right\| = \left(\int_{\mathbb{R}^3} 1 d\boldsymbol{x} \right)^{1/2} \to \infty. \tag{1.28}$$

Thus, the plane wave is not contained in the elements of the function space that quantum theory deals with, in the sense that it does not correspond to a vector of finite length. In mathematical terms, the plane wave is not an element of the function space $L^2(\mathbb{R}^3)$.

Another remarkable singularity of the plane wave is the continuity of its eigenvalues. In fact, since the plane wave $e^{i\boldsymbol{k}\cdot\boldsymbol{x}}$ is defined for any arbitrary $\boldsymbol{k} \in \mathbb{R}^3$, the eigenvalues $\hbar k_j$ belonging to the operator p_j take continuous values. Hence, the familiar technique of regarding the plane wave as a countably infinite base and resolving the function into the sum of independent components is not applicable. As a natural consequence, the momentum also cannot be expressed as a diagonal matrix. Again, the analogy with finite-dimensional linear algebra fails. However, instead of writing the function as the sum of independent components, it can be represented using integration. That is,

$$\psi(\boldsymbol{x},t) = \int \hat{\psi}(\boldsymbol{k},t) e^{i(\boldsymbol{k}\cdot\boldsymbol{x})} d\boldsymbol{k}. \tag{1.29}$$

This is nothing but the Fourier transform. If stated in this manner, the momentum operator can be expressed in a form similar to the proportionality rule of finitely many dimensions.

$$\begin{aligned} p_j u(\boldsymbol{x},t) &= -i\hbar \frac{\partial}{\partial x^j} u(\boldsymbol{x},t) \\ &= \hbar \int k_j \hat{u}(\boldsymbol{k},t) e^{i(\boldsymbol{k}\cdot\boldsymbol{x})} d\boldsymbol{k}. \end{aligned} \tag{1.30}$$

The continuous eigenvalues $\hbar k_j$ illustrated above are known in functional analysis as a continuous spectrum. The eigenfunctions $e^{i(\boldsymbol{k}\cdot\boldsymbol{x})}$ that belong to a continuous spectrum are singular eigenfunctions, unique functions that do not belong to the space. The same can also be said of the position operator. Since the eigenvalues related to the position also take continuous values, the eigenfunction of the position operator is also a singular eigenfunction. Therefore, to build quantum theory, it is necessary to consider the general eigenvalue problem containing such a continuous spectrum, and functional analysis plays a pivotal role in it.

Keypoint: Functional analysis provides a powerful tool to formulate counterintuitive concepts involved in the quantum theory.

Chapter 2

Topology

2.1 Fundamentals

2.1.1 What is topology?

Topology[1] plays a fundamental role in the study of infinite-dimensional spaces. This chapter explains what topology is and explores its usefulness in understanding the nature of infinite-dimensional spaces.[2]

"Topology" is a term that represents the geometric properties of point sets or spatial figures. It also refers to a branch of mathematics, which is often nicknamed "soft geometry." Topologists study the geometric properties or quantities that are conserved before and after deformation of an object, i.e., spatial extension or contraction. A simple example of the conserved property can be found in the deformation of a mug to a donut as depicted below.

Suppose we have a mug soft enough to be deformed at will. We could elongate the handle of the mug and contract its cup without cutting the mug or creating a fracture.[3]. The result of this would be a donut-shaped object. It should be noted that the mug and donut each have one hole, despite their different geometric appearances. This fact is expressed by saying that the one-hole property is conserved through the

---

[1] The word "topology" originates from the combination of two Greek words: topos (place) and logos (study).

[2] A complete discussion of general topology is beyond the scope of this chapter. Nevertheless, even minimal knowledge of topology provides a better understanding of the diverse subjects covered in functional analysis, such as the converging behavior of a point set, completion of a set, and the theory of wavelet transformation.

[3] The deformation of a mathematical object that is free from cutting or fracture is called a **continuous deformation**

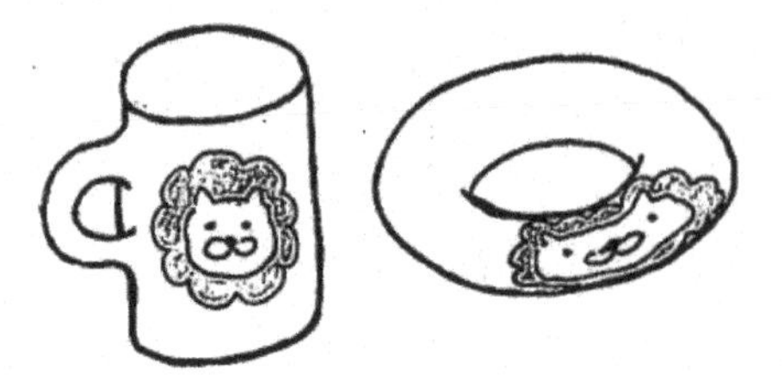

Figure 2.1: Deformation from a mug to a donut.

deformation from a mug to a donut (and vice versa). We say that the mug and donut have the same topology or that they are **topologically equivalent**.

Keypoint: Topology tells us what mathematical properties are conserved through deformation.

The above example seems trivial because it is easy to visualize deformation of three-dimensional spatial figures. On the other hand, functional analysis requires us to work with deformations of more abstract objects. For instance, transformations of point sets and sequences of functions are further from our natural intuition. Given this, how can we perceive the deformation of such abstract objects, and how can we describe it using the language of mathematics? To answer these questions, we must first look at the concept of closeness. Because there is neither "ruler" nor "measuring tape" available to evaluate the closeness between abstract objects, we need to invent mathematical concepts that enable us to measure the distance between two points in an abstract space.

2.1.2 What is closeness?

Let us imagine a notebook and pencil are lying on a desk. It is easy to determine which is closer to our hand. Or let us examine two routes to a service station. It is easy to say, "This route is much closer than that route." In both cases, we can derive the correct answer from a cursory inspection; hence, a discussion of the meaning of closeness is unnecessary.

Sometimes the criteria of closeness have nothing to do with spatial distance. For example, let us compare the era of Islam's ascendancy with the meridian of the Ottoman Empire. In the sense of elapsed years, the prime floruit of the Ottoman Empire is closer than the Islamic golden age to the present day. Similarly, let us imagine we are driving to a friend's house. In order to take the fastest route to our destination, we will consider several variables, and our determination of closeness will also include traffic and speed limits, not simply the total length of the route. Another example of closeness is suggested by a child's question: is an apple or banana closer to a vegetable? There is not a unique answer for this; the correct answer depends on the subjective views of those involved. This is because we have no universal criterion for closeness in the world of fruits and vegetables.

Figure 2.2: A grape and an orange; which is closer to a banana?

The absence of a universal criterion of closeness can also be found in functional analysis because there is no unique rule for determining the distance between mathematical entities such as numbers, vectors, and functions. Instead, we will establish a convenient and self-consistent rule for distance according to the nature of the sequences, sets, and spaces under consideration. Therefore, fundamental concepts such as continuity, convergence, and inclusion will have different meanings. When this happens, we say we are "introducing topology into the mathematical space (or set)."

> **Keypoint:** In functional analysis, we are allowed to set rules for distance between elements at our convenience.

2.1.3 Convergence of point sequences

The most fundamental concept, necessary for the study of topology, is that of **convergence of point sequences** in a given space. Before we continue, however, let us briefly review the precise meaning of distance between two points. Suppose a point P is located on the real number line, and assign the coordinate a to P, represented by $P(a)$. Then the distance between two points, $P(a)$ and $Q(b)$, is defined by

$$d(P,Q) = |b-a| \,. \tag{2.1}$$

This definition can be extended to a two-dimensional plane where the distance between $P(a_1,a_2)$ and $Q(b_1,b_2)$ is defined by

$$d(P,Q) = \sqrt{(b_1-a_1)^2+(b_2-a_2)^2} \tag{2.2}$$

using the **Pythagorean theorem**. Now we are ready to examine what is meant by an infinite sequence of points $P_1, P_2, \ldots, P_n, \ldots$ converging to P. It means:

$$\text{The distance } d(P_n, P) \text{ decreases unlimitedly with increasing } n. \tag{2.3}$$

This is symbolized by

$$\lim_{n\to\infty} P_n = P. \tag{2.4}$$

The plain statement given by (2.3) is commonly replaced by the following expression in mathematics:

$$\begin{gathered}\text{For any } \varepsilon > 0, \text{ there exists an integer } k \text{ such that} \\ \text{if } n > k, \text{ then } d(P_n, P) < \varepsilon.\end{gathered} \tag{2.5}$$

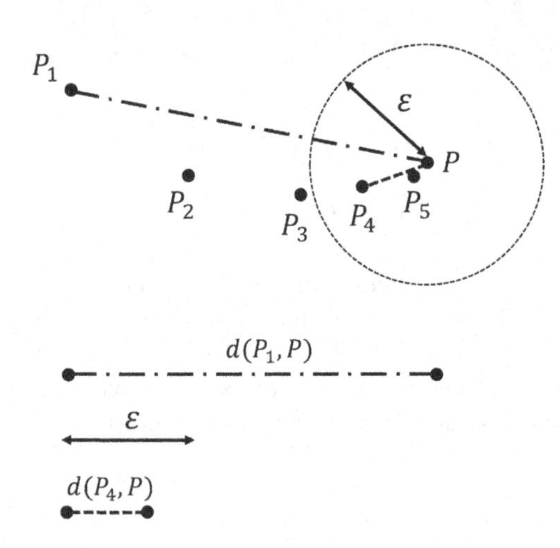

Figure 2.3: Converging behavior of a point sequence $\{P_n\}$ to the limit P. Even when a smaller ε is given, we can find an appropriate n that satisfies $d(P_n, P) < \varepsilon$.

The two statements, (2.3) and (2.5), differ in their vantage points or perspectives. In statement (2.3), the central figure is the infinite point sequence $P_1, P_2, \ldots, P_n, \ldots$. The statement (2.3) leads us to visualize a situation in which P_n approaches P as n increases, and the distance $d(P_n, P)$ decreases unlimitedly. On the other hand, the central figure in (2.5) is the point P. Here, we are standing at P as the observer and monitoring the small circular region centered at our feet with radius ε. Then we are observing that all points P_n for $n > k$ are contained in this circular region. In summary, the perspective in (2.3) moves ceaselessly with increasing n, but that in (2.5) is fixed at P.

In order to simplify the statement (2.5), we will examine the concept of a **neighborhood.**[4] An ε-neighborhood of P, $V_\varepsilon(P)$, is the set of points whose distances

[4] We will see later that the concept of neighborhood is heavily used in discussing the properties of topological spaces.

from the point P are smaller than ε, or

$$V_{\varepsilon}(P) = \{Q | d(P,Q) < \varepsilon\}. \tag{2.6}$$

Then the convergence of $\{P_n\}$ to P as $n \to \infty$ is written by:

For any $\varepsilon > 0$, there exists k such that
if $n > k$, then $P_n \in V_{\varepsilon}(P)$.

Figure 2.5 illustrates three different ways the sequence $\{P_n\}$ approaches P along a number line. We first assign the coordinates a_n and a to the points P_n and P, respectively, and then express the points by the symbols $P_n(a_n)$ and $P(a)$. In the top panel of Fig. 2.5, every a_n satisfies the relation

$$a_1 < a_2 < \cdots < a_n < \cdots < a. \tag{2.7}$$

Figure 2.4: The difference in the perspectives between the two statements: (2.3) and (2.5). The latter features a mouse that stays at P, while the former features a cat that approaches P.

In this case, we say that the convergent sequence $\{P_n\}$ is monotonically increasing. In contrast, the middle panel of Fig. 2.5 corresponds to the relation

$$a_1 > a_2 > \cdots > a_n > \cdots > a, \tag{2.8}$$

and we say that the convergent sequence $\{P_n\}$ is monotonically decreasing. The bottom panel of Fig. 2.5 gives a more general case in which the points P_n move to and fro while approaching the limit P. There are several other choices for the motion of P_n. It is possible to hybridize the three movements, or all the P_n for $n > k$ may be equivalent to each other so that $P_{k+1} = P_{k+2} = \cdots = P$.

Compared to the previous one-dimensional case, converging behavior of a point sequence $\{P_n\}$ in the two-dimensional plane is less straightforward, as can be seen in Fig. 2.6.

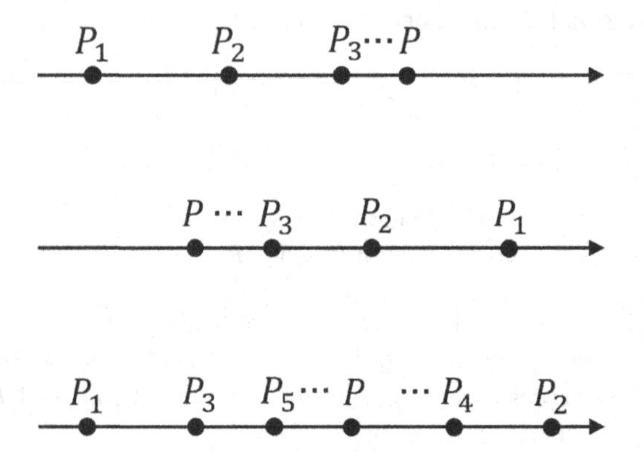

Figure 2.5: Three different ways of approaching the limit P.

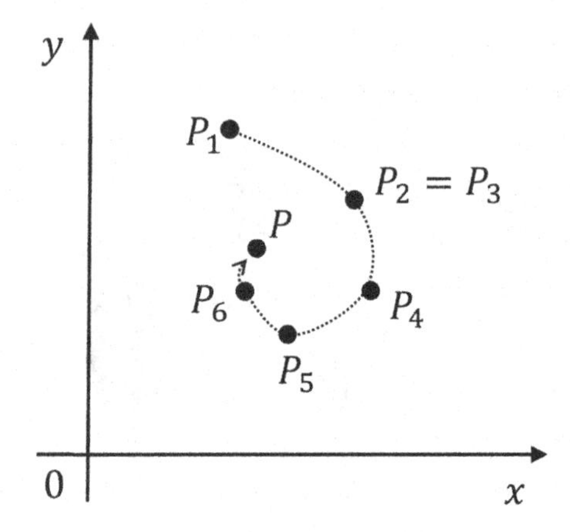

Figure 2.6: Convergence of a point sequence $\{P_n\}$ to P in a two-dimensional plane.

2.1.4 *Open set and closed set*

Suppose we have an infinite sequence of points P_n that converges to P, and assume all the elements $P_1, P_2, \cdots$ are contained in a set B. One important remark is that in this situation, the convergent point P may and may not be contained in B. Namely, the inclusion relationship between the infinite sequence $P_1, P_2, \cdots$ and a set B is independent of that between the convergent point P and set B. In-depth consideration on the distinction between these two relationships[5] leads us to the basic concepts: **open sets** and **closed sets**, as demonstrated below.

[5]For a finite sequence of points, the two relations are equivalent. Consider, for instance, a sequence consisting of ten points. If a given set contains all ten points, then it contains the convergent point (the tenth point), too.

Let us first consider a unit circle centered at the point P_0. We let B denote the interior of the circle (excluding points on the circumference) and $\overline{B}$ denote the union of B and all the points on the circumference. We call B and $\overline{B}$ an **open disk** and a **closed disk**, respectively. Figure 2.7 helps us visually distinguish between the two classes of disks. The circumference of the open disk is depicted by a dashed curve and that of the closed disk by a solid curve. The figure indicates that all the points on the circumference are excluded from or contained in the respective sets.

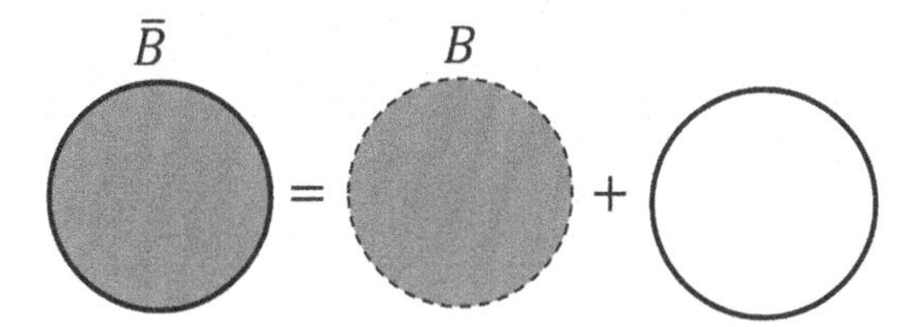

Figure 2.7: A closed disk $\overline{B}$ and an open disk B.

In terms of an ε-neighborhood, we can say that:

Theorem:

$$\text{Every point } P \text{ in an open set } B \text{ satisfies } V_\varepsilon(P) \subset B. \quad \blacksquare \tag{2.9}$$

Here, the constant ε is assumed to be positive and small enough.

It should be noted that in the closed disk $\overline{B}$, there is a point that does not satisfy this condition. An obvious example is a point on the circumference. If we choose this point as P, then $V_\varepsilon(P)$ must contain a point outside the circumference no matter how small ε is. Therefore, the condition $V_\varepsilon(P) \subset \overline{B}$ fails. This fact is visualized in Fig. 2.8.

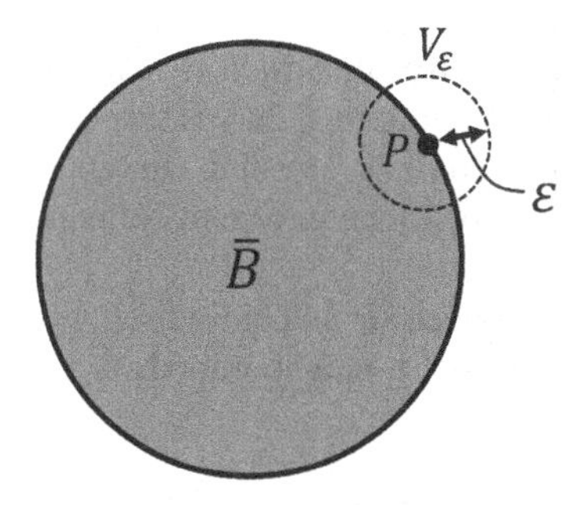

Figure 2.8: Schematics for that $V_\varepsilon(P) \subset \overline{B}$ fails in a closed disk $\overline{B}$.

Next, we will look into the following property of a closed disk:

$$\text{If a point sequence } \{P_n\} \text{ in a closed disk } \overline{B} \text{ converges to } P, \text{ then } P \in \overline{B}. \tag{2.10}$$

This assertion is not true for an open disk, B. For example, suppose that $\{P_n\}$ in B converges to P, which is located on the circumference. In this case, $P_n \in B$ for any n but $P \notin B$.

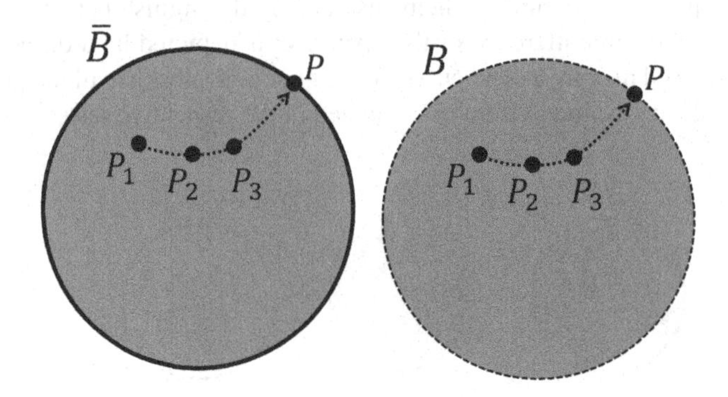

Figure 2.9: Left: A closed disk $\overline{B}$ always contains the limit P of $\{P_n\}$. Right: An open disk B may not contain the limit P of $\{P_n\}$.

> **Keypoint:** Even if all the points $\{P_n\}$ are contained in a set X, their limit P may not be contained in X.

The discussion on open/closed sets leads us to a broader study of general sets of points. Figure 2.10 gives two examples of point sets whose geometries differ from the unit disk. The set in Fig. 2.10(a) has a donut-like shape, excluding the points on the dashed curves (inner and outer boundaries). This donut-like set allows the statement (2.9) to hold similarly to the open-disk case; the set depicted in Fig. 2.10(a) contains a neighborhood of every point in the set. As a consequence, we say that this set is an open set.

Figure 2.10(b) shows an indecisive set, which is neither open nor closed.[6] The upper half of the boundary consists of dashed curves that comply with the statement (2.9), whereas the bottom half consists of solid curves complying with (2.10). Therefore, the set in Fig. 2.10(b) does not conform to only one of the two statements, (2.9) or (2.10).

We are now ready to define an open set and closed set, as generalizations of open and closed disks, using the concept of ε-neighborhoods.

Definition (Open set):
Suppose that arbitrary point P in a set S satisfies $V_\varepsilon(P) \subset S$ for a sufficiently small ε. Then S is called an **open set**.[7] ∎

[6] It is a common misunderstanding that all sets are either open or closed. In reality, we can easily find many sets that are neither open nor closed.

[7] When an open set is a subset of the real number line, it is called an **open interval**.

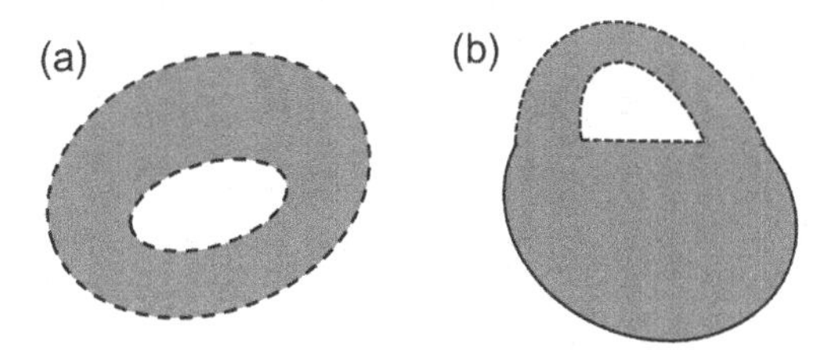

Figure 2.10: (a) An open set with a donut-like shape. (b) A set which is neither open nor closed.

Definition (Closed set):
Suppose a set S contains an infinite sequence of points P_n that converges to P. If $P \in S$, then S is called a **closed set.**[8]. ■

For convenience, we stipulate that the **empty set** $\varnothing$ is both open and closed, although $\varnothing$ contains no points and neither of the two definitions makes sense.

2.1.5 Accumulation points

In our discussion of the infinite point sequence $P_1, P_2, \ldots, P_n, \ldots$ that approaches the convergent point P, we have not mentioned the special situation expressed by $P_1 = P_2 = \cdots = P$. In the latter case, the points P_n are identical, fixed at position P. Although this sequence converges to P, it is more natural to consider the condition in which the elements $P_1, P_2, \ldots, P_n, \ldots$ are distinct from one another and become dense when approaching P. Under this condition, the convergent point is specially defined.

Definition (Accumulation point):
Suppose a set M in the plane (or on a line) contains an infinite sequence $\{P_n\}$ that consists of different points. If $\{P_n\}$ converges to P, then P is called an accumulation point[9] of M. ■

It follows from the definition that no accumulation point exists when a set M consists of a finite number of points. It is also understood that when M is closed, all the accumulation points of M are contained in M. However, when M is open, its accumulation points are not always contained in M, as seen in the example below.

Example 1:
Every point x in the set $\mathbb{R}$ is an accumulation point of the set $\mathbb{Q}$. For any $\varepsilon > 0$ and any real number a, we can find infinitely many rational numbers x that satisfy $|x - a| < \varepsilon$. ■

---

[8] When a closed set is a subset of the real number line, it is called a **closed interval**

[9] In general, P may or may not be contained in M.

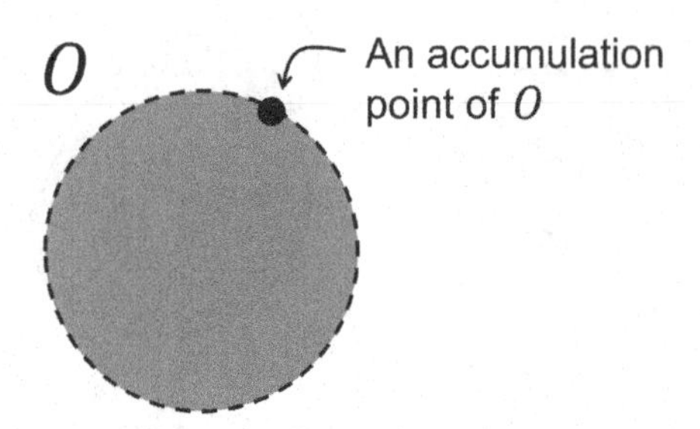

Figure 2.11: A point just on the outer boundary of an open disk O is an accumulation point of O.

Example 2:
Suppose a set M on the number line is defined by

$$\begin{aligned} M &= \left\{0, \frac{1}{2}, -\frac{2}{3}, \frac{3}{4}, -\frac{4}{5}, \frac{5}{6}, -\frac{6}{7}, \ldots\right\} \\ &= \left\{(-1)^n \left(1 - \frac{1}{n}\right) \middle| \, n = 1, 2, \ldots\right\}. \end{aligned} \tag{2.11}$$

We can see that M has two accumulation points: 1 and -1. This is clear if we look at the elements one by one, alternating them and lining them up:

$$\frac{1}{2}, \frac{3}{4}, \frac{5}{6}, \ldots \to 1 \tag{2.12}$$

$$0, -\frac{2}{3}, -\frac{4}{5}, -\frac{6}{7}, \ldots \to -1. \tag{2.13}$$

We should also realize that the two accumulations 1 and -1 do not belong to M. ■

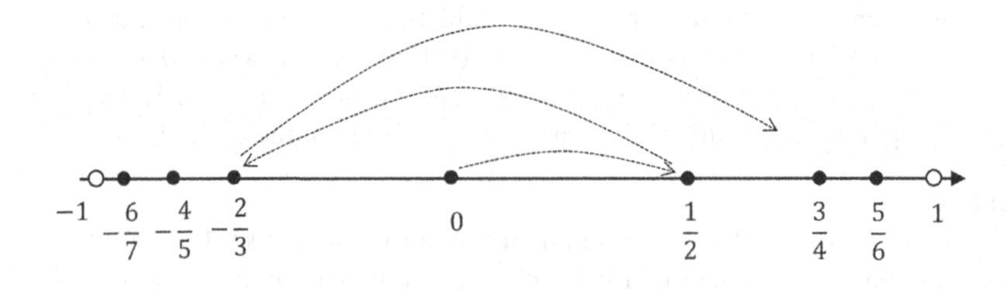

Figure 2.12: A point sequence that come and go across the origin 0 and eventually approaches the two accumulation points: -1 and $+1$.

Example 3:
Suppose an open set O in the coordinate plane is defined by[10]

$$O = \{ (x,y) \mid x^2 + y^2 < 1 \}. \tag{2.14}$$

Then accumulation points of O are all the points contained in O and the points just on the circumference of the circle are defined by $x^2 + y^2 = 1$. ■

Keypoint:
1. Given an accumulation point P, you can find infinitely many points within a small region around P, no matter how small the region is.

2. An accumulation point of a set X may and may not be contained in X.

Given an open set O, every point of O is an accumulation point[11] of O. This is because every point of O is surrounded by an infinitely large number of points, all of which are contained in O. In other words, an infinite number of points are accumulated close to any point of O. However, the opposite is not true. An accumulation point of O is not always a point in O.

Let us expand our discussion to more general sets M which are neither open nor closed. Even in this situation, almost all points in M should be accumulation points of M. But we must take care that "almost all" is not "all." The exceptions arise if M contains **isolated points**. Indeed, any isolated point in M cannot serve as an accumulation point of M.

For a better understanding, we will look into the necessary and sufficient condition for a point P of M *not* to be an accumulation point of M. The condition is written as

$$V_\varepsilon(P) \cap M = \{P\}, \tag{2.15}$$

where ε is sufficiently small. Figure 2.13 gives a schematic of the relationship between P, $V_\varepsilon(P)$, and M represented by (2.15). Plainly speaking, it describes the situation in which no point exists in the vicinity of P. Hence, P looks like a tiny, isolated island far from the mainland M, although P belongs to the territory of M. In summary, there are only two possibilities for every point contained in M: it is an isolated point or it is an accumulation point of M.

---

[10]The set O is an open set surrounded by the unit circle and centered at the origin. Any point on the circumference of the circle is not contained in O.

[11]The only exception is the empty set $\varnothing$. There is no point in $\varnothing$; therefore, $\varnothing$ possesses no accumulation point, although it should be an open set.

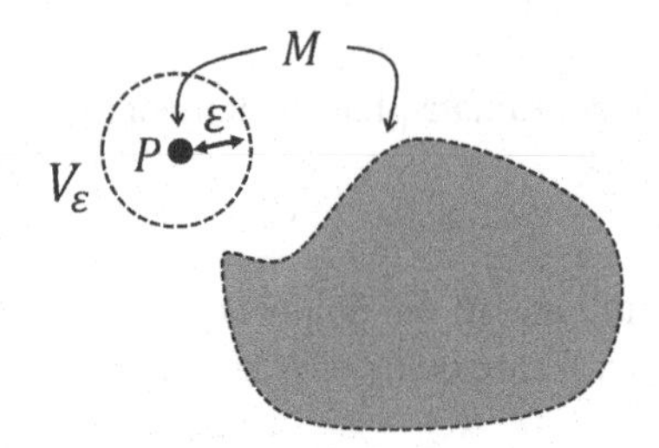

Figure 2.13: Although the point P is a member of the set M, the P is not an accumulation point of M.

2.2 Continuous mapping

2.2.1 *Relation between continuity and open sets*

Consider a real-valued function $y = f(x)$ that maps a real number x to another real number y. It may or may not be continuous, but we can evaluate its continuity using elementary mathematical analysis (see Appendix). The concept of continuity is applicable to broader mathematical entities beyond the continuity of functions relating real numbers. All we need is the ability to define distance between entities and convergence of an infinite sequence. An infinite point sequence in the two-dimensional plane is an example of this, so let us consider the concept of continuity in a mapping between two-dimensional planes.

Definition (Continuity of mapping at a point):
Suppose that a mapping φ between two planes satisfies the following condition for arbitrary points in the planes:

$$P_n \to P \; [n \to \infty] \Rightarrow \varphi(P_n) \to \varphi(P) \; [n \to \infty] \tag{2.16}$$

Then we say φ is continuous at P. ■

In simple terms, φ is continuous at P if the convergence property of infinite sequences is conserved at P. If φ is discontinuous, the image $\{\varphi(P_n)\}$ may not converge. From a geometric viewpoint, continuity of the mapping φ is a sort of mechanical deformation of an elastic object without tearing or making holes. The deformation from a mug to a donut is a typical example of a mapping that is continuous everywhere. If we create a crack or hole in the mug during the deformation, then the mapping loses continuity at the failure because the image $\{\varphi(P_n)\}$ no longer converges in the vicinity of the crack or hole.

We can restate the converging behaviors of $P_n \to P$ and $\varphi(P_n) \to \varphi(P)$ using the concept of neighborhood.

1. The proposition $P_n \to P$ with $n \to \infty$ is equivalent to the statement:

$$\begin{gathered}\text{Given } \delta > 0 \text{, there exists an integer } k \text{ such that} \\ n > k \Rightarrow P_n \in V_\delta(P).\end{gathered} \tag{2.17}$$

2. The proposition $\varphi(P_n) \to \varphi(P)$ with $n \to \infty$ is equivalent to the statement:

$$\begin{aligned} &\text{Given } \varepsilon > 0, \text{ there exists an integer } k' \text{ such that} \\ &n > k' \Rightarrow \varphi(P_n) \in V_\varepsilon(\varphi(P)). \end{aligned} \tag{2.18}$$

By substituting the statements (2.17) and (2.18) into (2.16), we can express the continuity of a mapping with respect to the inclusive relationship between neighborhoods. In fact, the original definition (2.16) is equivalent to the one below.

Theorem:
A mapping φ is continuous at P if the following condition is satisfied:

$$\begin{aligned} &\text{Given } \varepsilon > 0, \text{ there exists } \delta > 0 \text{ such that} \\ &Q \in V_\delta(P) \Rightarrow \varphi(Q) \in V_\varepsilon(\varphi(P)). \quad \blacksquare \end{aligned}$$

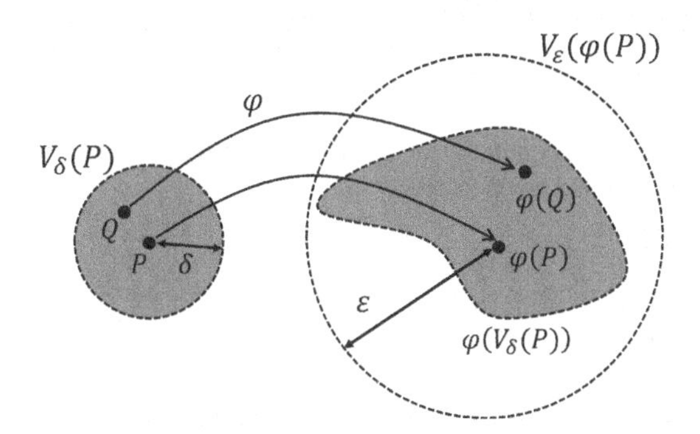

Figure 2.14: Continuous mapping φ transforms any point that close to P into a point that close enough to $\varphi(P)$.

It should be emphasized that the point Q can be arbitrarily chosen in the set $V_\delta(P)$. This leads us to an important consequence.

Theorem:
A mapping φ is continuous at P if the following condition is satisfied:

$$\begin{aligned} &\text{Given } \varepsilon > 0, \text{ there exists } \delta > 0 \text{ such that} \\ &\varphi(V_\delta(P)) \subset V_\varepsilon(\varphi(P)). \quad \blacksquare \end{aligned} \tag{2.19}$$

In the discussion so far, we have focused on the continuity of the mapping φ at a particular point P. It leads us to question whether there is a condition such that φ would be continuous everywhere[12] in a given set. If so, can we formulate this condition in a simple manner? The answer to both of these questions is yes.

[12]When the continuity holds everywhere, φ is called a **continuous mapping**. This definition is similar to that of a **continuous function**. When a function $f(x)$ is continuous at every x, we call it a continuous function.

Theorem (Continuous mappings and open sets):
A mapping φ is continuous everywhere if and only if the following condition is satisfied:

Given an arbitrary open set O,
its inverse image $\varphi^{-1}(O)$ is also an open set. ■

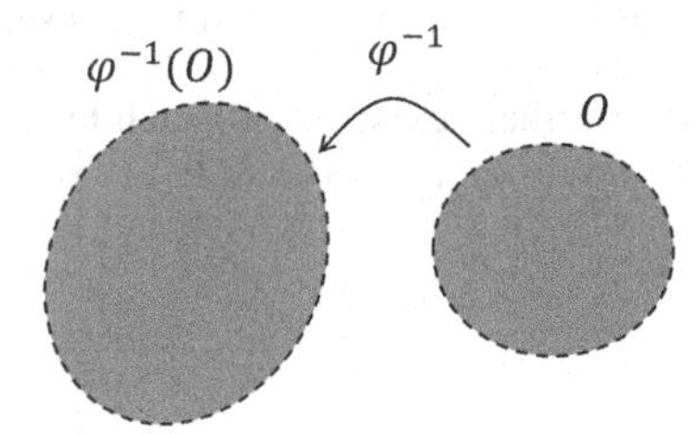

Figure 2.15: Criterion for the continuity of φ. If it is continuous, then the inverse image $\varphi^{-1}(O)$ of an open set O should be open, too.

Proof:

1. We will first prove the necessity that

$$\varphi \text{ is continuous} \Rightarrow \varphi^{-1}(O) \text{ is open for arbitrary open set } O.$$

Suppose we have a continuous mapping φ and an open set O, then $\varphi^{-1}(O)$ is the set created by the inverse mapping φ^{-1}. We choose an arbitrary point $P \in \varphi^{-1}(O)$. The image of P, $\varphi(P)$, is obviously contained in O, and we can say $\varphi(P) \in O$. Since O is open, we have

$$V_\varepsilon(\varphi(P)) \subset O \tag{2.20}$$

for ε small enough. Also, since φ is continuous at P, we have [See (2.19)]

$$\varphi(V_\delta(P)) \subset V_\varepsilon(\varphi(P)) \tag{2.21}$$

for sufficiently small δ. The two relations (2.20) and (2.21) imply that

$$\varphi(V_\delta(P)) \subset O, \tag{2.22}$$

or equivalently that

$$V_\delta(P) \subset \varphi^{-1}(O). \tag{2.23}$$

From our assumption, the last inclusive relation must hold for any arbitrary P in $\varphi^{-1}(O)$. This means that $\varphi^{-1}(O)$ is an open set.

2. Next we will address the sufficiency that

$$\varphi^{-1}(O) \text{ is open for arbitrary open set } O \Rightarrow \varphi \text{ is continuous.}$$

Given an open set, we know its inverse image under φ^{-1} is open, too, by assumption. Our task is to examine whether the statement (2.19) is true at an arbitrary point P in the set that is to be transformed through φ. We first note that an ε-neighborhood around the point $\varphi(P)$ is an open set. We defined an open set O by

$$O = V_\varepsilon(\varphi(P)) \tag{2.24}$$

where P is arbitrarily chosen. It is obvious that $P \in \varphi^{-1}(O)$, and from the assumption $\varphi^{-1}(O)$ is an open set. Therefore, we can choose a positive constant δ that is small enough to satisfy the inclusive relation $V_\delta(P) \subset \varphi^{-1}(O)$.

This is equivalent to the relation

$$\varphi(V_\delta(P)) \subset O. \tag{2.25}$$

We rewrite O using (2.24) to obtain

$$\varphi(V_\delta(P)) \subset V_\varepsilon(\varphi(P)). \tag{2.26}$$

Comparing this result with the definition of continuity given by (2.19), we conclude that φ is continuous at P. Finally, because P was arbitrary chosen, φ is continuous everywhere. This means that φ is a continuous mapping. ■

2.2.2 Relation between continuity and closed sets

In the preceding section, we learned that φ is a continuous mapping if and only if $\varphi^{-1}(O)$ is open for an arbitrary open set O. This naturally leads us to examine the counterpart for a closed set.

Theorem (Continuous mappings and closed sets):
A mapping φ is continuous everywhere if and only if, given an arbitrary closed set F, its inverse image $\varphi^{-1}(F)$ is also closed. ■

Keypoint: The continuity of a mapping φ is judged by the nature of its inverse φ^{-1}.

We should remark that the continuity of a mapping φ is determined not by the property of its images, $\varphi(O)$ or $\varphi(F)$, but the property of its inverse images, $\varphi^{-1}(O)$ and $\varphi^{-1}(F)$. In fact, even when $\varphi(O)$ is open, it tells us nothing about the continuity

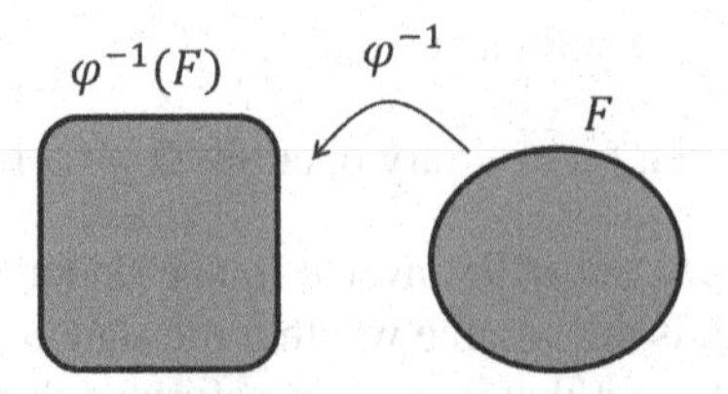

Figure 2.16: Alternative criterion for the continuity of φ. If it is continuous, then the inverse image $\varphi^{-1}(F)$ of a closed set F should be closed, too.

of φ. Similarly, whether or not $\varphi(F)$ is closed is irrelevant to the continuity of φ. The following three examples will help clarify our meaning.

Example 1:
Consider a continuous mapping (continuous function)

$$y = x^2, \tag{2.27}$$

which maps an open interval $(-1,1)$ on the x axis to $[0,1)$ on the y axis. In this case, the image $[0,1)$ is neither open nor closed. ■

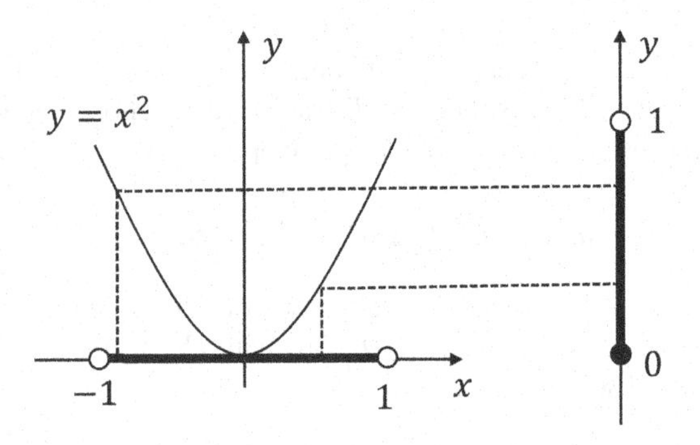

Figure 2.17: The mapping $f : x \to x^2$ on $(-1,1)$ is continuous, while the image of the open interval $(-1,1)$ is neither open nor closed.

Example 2:
Suppose we have a continuous mapping φ between two-dimensional planes defined by

$$\varphi(x,y) = \begin{cases} (x,y), & y \geq 0 \\ (x,-y), & y < 0 \end{cases}. \tag{2.28}$$

Figure 2.18 gives a diagram of the mapping. Here, O is an open set consisting of internal points of the unit circle around the origin. It is transformed into $\varphi(O)$, the upper half of the unit circle. Note that $\varphi(O)$ is not an open set because it includes the points along the x axis. ■

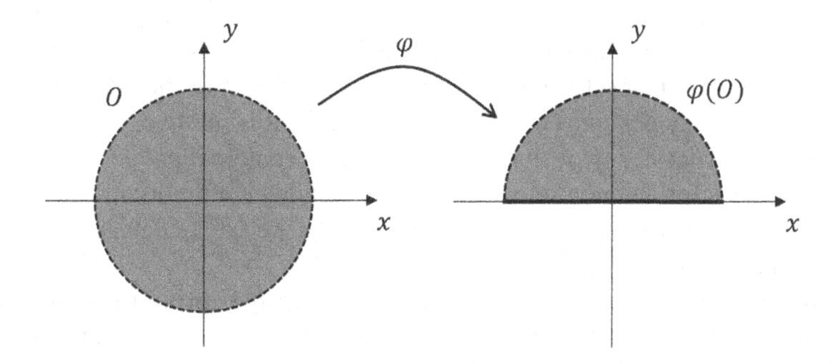

Figure 2.18: The mapping $\varphi(x, y)$ given by (2.28) is continuous, while the image of an open unit circle is neither open nor closed.

Example 3:
Consider the continuous function defined by

$$y = \tan^{-1} x. \tag{2.29}$$

The closed interval $[0, \infty)$ on the x axis is mapped to $[0, 1)$ on the y axis, which is not a closed set. ■

So far we have learned that the presence/absence of continuity in a mapping φ is determined by its inverse image of open or closed sets. It should be emphasized that the concept of openness is based on the global, not local, characteristics of a set. Hence, when evaluating the continuity of φ on a set M, we does not need to check convergence at certain points in M. This contrasts the continuity in functions studied in elementary mathematical analysis.

Keypoint: The inverse φ^{-1} is what we should pay attention to in examining the continuity of φ.

When evaluating the continuity of a function $y = f(x)$, we pay attention to its continuity at a particular point P on the x axis. Then we check to see whether the local continuity is found at arbitrary points along the x axis. If it is, then the function is globally continuous, and we call it a continuous function. This procedure for evaluating continuity seems reasonable in the realm of elementary analysis because function continuity produces two properties. First, two neighboring points before a mapping remain close to each other after the mapping. This is called **pointwise continuity**. Second, this conservation in the closeness holds true at arbitrary positions. This is called **uniform continuity**.

Importantly, the definition of uniform continuity based on an open set seems independent of the local properties of the set, such as pointwise continuity. Instead, the mixture of local and global behaviors of the set is inherent in the concept of open sets. Uniform continuity implies global properties in the sense that it describes the properties of a subset surrounded by a certain boundary. It also implies local properties because an open set O must satisfy the requirement of the neighborhood: $V_\varepsilon(P) \subset O$ for arbitrary positive constant ε. In summary, it is fair to say that an open set is a global entity defined by its local condition. This combination of the local and global requirements for an open set is suitable for defining continuity in a mapping. Two requirements, pointwise continuity and uniformity, are satisfied in an artful way by a single concept: the open set.

Before moving on, let us take a look to the concept of **discontinuous mappings**. When a mapping φ is discontinuous, then its inverse image $\varphi^{-1}(O)$ of an open set O is not open. Below is a simple example of this.

Example 1:
Suppose we have a function

$$\varphi(x) = \begin{cases} x, & x \leq 0, \\ x+2, & x > 0, \end{cases} \tag{2.30}$$

which is discontinuous at $x = 0$. The function satisfies the relation

$$\varphi^{-1}((-1,1)) = (-1,0], \tag{2.31}$$

where the inverse image $(-1,0]$ is not open even though $(-1,1)$ is open. ■

Example 2:
The discontinuous function

$$\varphi(x) = \begin{cases} 1, & x \text{ is irrational}, \\ 0, & x \text{ is rational}, \end{cases} \tag{2.32}$$

satisfies the relation

$$\varphi^{-1}\left(\left(-\frac{1}{2}, \frac{1}{2}\right)\right) = \text{a set of rational numbers}. \tag{2.33}$$

Although $\left(-\frac{1}{2}, \frac{1}{2}\right)$ is open, the set of rational numbers is not open. ■

2.2.3 *Closure of a set*

In the previous section, we looked at the relationship between continuous mappings and open sets. Now we will examine another important concept in mapping continuity, called closure.

Definition (Closure):
The closure of a set S, denoted $\overline{S}$, is the union of S and all its accumulation points. ■

The above definition is equivalent to[13]

$$\overline{S} = S \cup \{\text{All accumulation points of } S\}. \tag{2.34}$$

Let us look at a few simple examples of closure.

Example 1:
Consider a set of points S defined by

$$S = \left\{1, \frac{1}{2}, \frac{1}{3}, \ldots, \frac{1}{n}, \ldots\right\}. \tag{2.35}$$

It possesses only one accumulation point, 0, and we have

$$\overline{S} = \left\{1, \frac{1}{2}, \frac{1}{3}, \ldots, \frac{1}{n}, \ldots, 0\right\}. \quad \blacksquare \tag{2.36}$$

Example 2:
Consider a set of points S on the two-dimensional plane defined by

$$S = \{\,(x,y) \mid 0 < x^2 + y^2 < 1\}. \tag{2.37}$$

All points (x,y) that satisfy this condition are inside the unit circle except for its center, $(0,0)$. Therefore, the closure $\overline{S}$ is the union of S and the origin and the points on the circle as expressed by

$$\overline{S} = \{\,(x,y) \mid 0 \le x^2 + y^2 \le 1\}. \quad \blacksquare \tag{2.38}$$

Now we are ready to relate the continuity in the mapping φ to the concept of closure.

Theorem (Continuous mapping and closure):

A mapping φ is continuous in $X \Leftrightarrow$ All subsets S in X satisfy $\varphi\left(\overline{S}\right) \subset \overline{\varphi(S)}$. $\blacksquare$

[13] The only exception is the empty set $\varnothing$, which is an open set for convenience. We define $\overline{\varnothing} = \varnothing$, even though $\varnothing$ does not contain any points.

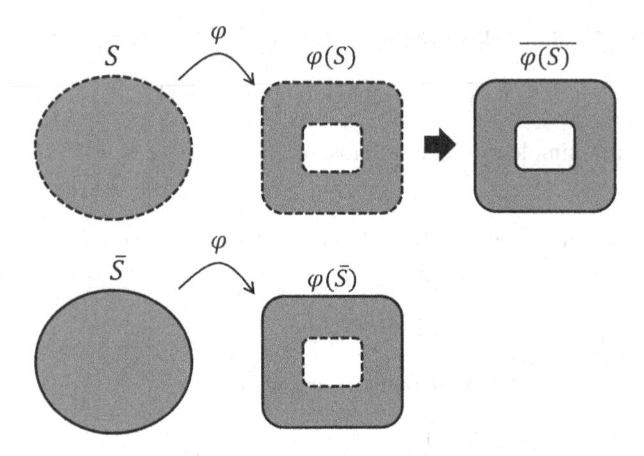

Figure 2.19: Given a continuous mapping φ, the image $\varphi\left(\overline{S}\right)$ is necessarily included in $\overline{\varphi(S)}$.

Proof:

1. We will first address the necessity,

$$\text{“}\varphi \text{ is continuous”} \Rightarrow \varphi\left(\overline{S}\right) \subset \overline{\varphi(S)}. \tag{2.39}$$

   It suffices to show that, for a continuous mapping φ, an arbitrary point $x \in \overline{S}$ is mapped to $\varphi(x) \in \overline{\varphi(S)}$. This trivially holds for $x \in S$. Therefore, we will restrict our discussion to the case where $x \notin S$ but x is an accumulation point of S.
   Consider an infinite sequence of points $x_1, x_2, \ldots, x_n, \ldots$, all of which are contained in S. Also assume that $x_n \to x \notin S$. The continuity of φ implies that the sequence $\varphi\left(x_n\right)$ converges to $\varphi(x)$, since $x_n \to x$. It is obvious that $\varphi\left(x_n\right) \in \varphi(S)$ for every n and that $\varphi(S) \subset \overline{\varphi(S)}$. Thus, $\varphi(x) \in \overline{\varphi(S)}$.
2. Next we consider the conditional statement:

$$\varphi\left(\overline{S}\right) \subset \overline{\varphi(S)} \Rightarrow \text{“}\varphi \text{ is continuous.”} \tag{2.40}$$

   Below we use a proof of contradiction. Suppose that $x_n \to x$ holds but $\varphi\left(x_n\right) \to \varphi(x)$ does not hold. Under these conditions, we can pick up a subset of an infinite sequence $\{x_{k_1}, x_{k_2}, \ldots, x_{k_i}, \ldots\}$ from the original infinite sequence $\{x_1, x_2, \cdots, x_n, \cdots\}$ so that its image $\varphi\left(x_{k_i}\right)$ satisfies

$$d\Big(\varphi\left(x_{k_i}\right), \varphi(x)\Big) \geq \varepsilon \quad [i = 1, 2, \ldots] \tag{2.41}$$

   for an arbitrary positive constant ε. Using the subset, we define a set S by

$$S = \left\{x_{k_1}, x_{k_2}, \ldots x_{k_i}, \ldots\right\}. \tag{2.42}$$

The accumulation point of S is x. Hence, we have

$$\overline{S} = \left\{x_{k_1}, x_{k_2}, \dots x_{k_i}, \dots, x\right\}. \tag{2.43}$$

This means that the point x is contained in $\overline{S}$.

Now let us consider the property of the set $\varphi(S)$ defined by

$$\overline{\varphi(S)} = \overline{\left\{\varphi\left(x_{k_1}\right), \varphi\left(x_{k_2}\right), \cdots, \varphi\left(x_{k_i}\right), \cdots\right\}}. \tag{2.44}$$

Note that the point $\varphi(x)$ is not contained in the set $\overline{\varphi(S)}$ because $\varphi(x_{k_i})$ does not converge to $\varphi(x)$. In summary, we have $x \in \overline{S}$ but $\varphi(x) \notin \overline{\varphi(S)}$, which implies $\varphi\left(\overline{S}\right) \not\subset \overline{\varphi(S)}$. However, this is contradiction to our initial hypothesis that $\varphi\left(\overline{S}\right) \subset \overline{\varphi(S)}$. Therefore, the sequence $\varphi\left(x_n\right)$ must converge to $\varphi(x)$, which guarantees the continuity of the mapping φ. ■

It is important to note that the concept of point sequence convergence is unnecessary for testing continuity in the mapping φ. On the other hand, the concept of closure plays a key role in determining the continuity of φ, although it appears irrelevant at a glance.

Why would we use the concept of closure to define a continuous mapping? This question leads us to a discussion of homeomorphisms. Plainly speaking, a homeomorphism is something like a deformation from a mug to a donut. The deformation does not preserve the point-to-point distance; however, it preserves the openness and closeness of a set, the neighborhood of a point, closure of a set, and many more topological properties. After discussing the basics of topology, it becomes convenient to express continuous mappings using the language of topological properties such as closure.

2.3 Homeomorphism

2.3.1 Homeomorphic mapping

When the mapping φ is continuous, its inverse φ^{-1} may or may not be continuous. Particularly when φ^{-1} is also continuous, then it is called to be **homeomorphic** as defined below.

Definition (Homeomorphic mapping):
Consider a continuous mapping φ which transforms the metric space[14] (X, d) to another metric space (Y, d'). If φ^{-1} is also continuous, we say that:
1. X and Y are homeomorphic (or topologically equivalent)
2. φ is a homeomorphic mapping (or homeomorphism) from X to Y. ■

[14] The notation (X, d) indicates that the metric space X is endowed with the distance d between two points. We have many choices for the definition of d in X, although the most conventional one is based on the Pythagorean theorem; See (2.45).

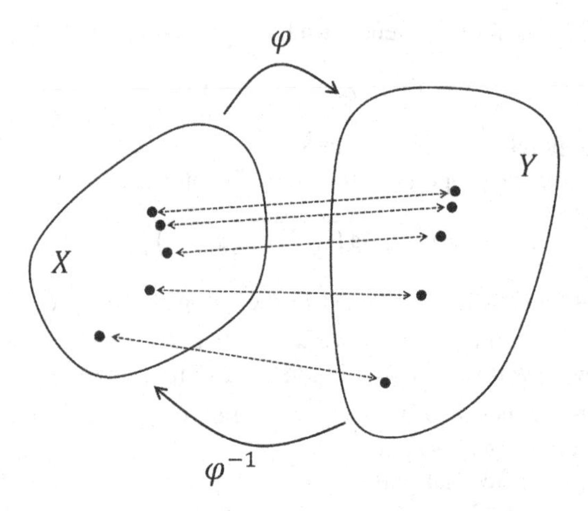

Figure 2.20: Homeomorphic mapping φ from X to Y.

Emphasis should be put on the fact that, in general, homeomorphic mappings do not conserve the distance between two points. For example, consider the two-dimensional Euclidean space $\mathbb{R}^2$. The distance $d(x,y)$ from the point x to y in $\mathbb{R}^2$ is usually defined by

$$d(x,y) = \sqrt{(x_1 - y_1)^2 + (x_2 - y_2)^2}. \tag{2.45}$$

However, this definition is not unique. There are many other choices for the definition of the point-to-point distance in $\mathbb{R}^2$. For instance, we can define the distance d' between the points x and y in $\mathbb{R}^2$ by

$$d'(x,y) = |x_1 - y_1| + |x_2 - y_2|. \tag{2.46}$$

The value of the distance between x and y in $\mathbb{R}^2$ depends on the way we define the point-to-point distance. This is the reason a homeomorphic mapping φ from $(\mathbb{R}^2, d)$ to $(\mathbb{R}^2, d')$ does not conserve the distance between points. On the other hand, no matter how we define the distance, the convergence property of a point sequence $\{x_n\}$ in $\mathbb{R}^2$ remains unchanged through the homeomorphic mapping φ. Meaning, if $\{x_n\}$ converges to x under the defined distance d, it converges to x under d', too, and vice versa. This is one of the most fundamental features of homeomorphic mappings: they do not conserve the distance between points, but they do conserve the convergence properties of the point sequences in the given space.

Keypoint: Homeomorphic mappings are special continuous mappings that preserve the convergence (not distance) property.

Besides the convergence property, homeomorphic mappings conserve several other properties of the spaces being mapped. All of them are associated with the topology of the space.

Theorem:
Consider a homeomorphic mapping φ from a metric space (X,d) to (Y,d'). Then we have:
1. $x_n \to x \Leftrightarrow \varphi(x_n) \to \varphi(x)$.
2. O is an open set in $X \Leftrightarrow \varphi(O)$ is an open set in Y.
3. F is a closed set in $X \Leftrightarrow \varphi(F)$ is a closed set in Y.
4. V is a neighborhood of $x \in X \Leftrightarrow \varphi(V)$ is a neighborhood of $\varphi(x) \in Y$.
5. The closure of a set S in X, $\overline{S}$, is mapped to the closure of $\varphi(S)$ in Y, $\overline{\varphi(S)}$. ■

For statements 1 through 4, we can prove the conditional statements in the direction $\Longrightarrow$ using the continuity of the inverse mapping φ^{-1}. Conversely, the statements in the direction $\Longleftarrow$ are due to the continuity of φ.

In order to justify 5, it is helpful to remember that

$$\overline{\varphi(S)} = \varphi\left(\overline{S}\right) \tag{2.47}$$

In fact, the continuity of φ implies

$$\varphi\left(\overline{S}\right) \subset \overline{\varphi(S)}, \tag{2.48}$$

and the continuity of φ^{-1} implies

$$\varphi^{-1}\left(\overline{\varphi(S)}\right) \subset \overline{\varphi^{-1}\left(\varphi(S)\right)} = \overline{S}, \tag{2.49}$$

from which we obtain

$$\overline{\varphi(S)} \subset \varphi\left(\overline{S}\right) \tag{2.50}$$

Comparing the results of (2.48) and (2.50), we can conclude (2.47).

2.3.2 *Revisited: What is topology?*

Throughout this chapter, we have learned the basic concepts that will allow us to define the topology of a space as stated below.

Definition (Topological equivalence):
Two metric spaces (X,d) and (Y,d') are said to have the **same topology**,[15] if they are **homeomorphic**. ■

The topology of a space is the collection of all the properties conserved under a homeomorphic mapping. If a mapping is homeomorphic, then it conserves the

[15] We can restate this as "they are topologically equivalent."

convergence of point sequences, the openness and closeness of a set, the closure property of a set, and the neighborhood of a point. All of the properties that determine the "closeness" between points remain unchanged under the homeomorphic mapping. These properties are what we call the topological properties of the space.

> **Keypoint:** Topological properties of a space are what are conserved by homeomorphic mappings.

The mathematical concept of closeness is not always described by the distance between entities. Therefore, we need to introduce concepts that will enable us to characterize (or measure) closeness, such as with the concepts of open sets, closed sets, and closure.

Example 1:
The open interval $(-1,1)$ is endowed with the same topology as that of the set of all real numbers $\mathbb{R}$. In fact, this interval is transformed to $\mathbb{R}$ via the homeomorphic mapping φ defined by[16]

$$\varphi : x \in (-1,1) \to \tan\left(\frac{\pi}{2}x\right) \in \mathbb{R} \quad \blacksquare \tag{2.51}$$

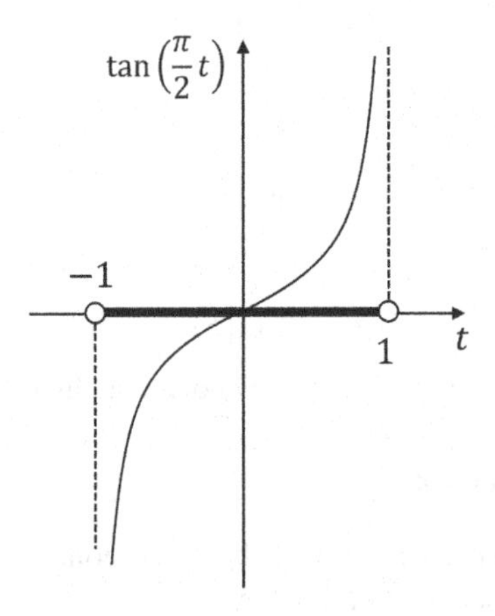

Figure 2.21: Homeomorphic mapping φ from $(-1,1)$ to $\tan\left(\frac{\pi}{2}x\right)$.

[16]Generally, there are many kinds of homeomorphic mappings that relate two spaces with the same topology. For instance, the mapping defined by $\varphi : x \to \tan\left(\frac{\pi}{2}x^{2n+1}\right)$ $(n = 1,2,3,\ldots)$ also satisfies the conditions.

Example 2:
The inside of the unit circle $B = \{ (x,y) \,|\, x^2+y^2 < 1 \}$ and the plane $\mathbb{R}^2$ are topologically equivalent. An example of a homeomorphic mapping that relates the two spaces is

$$\varphi : (r\cos\theta, r\sin\theta) \in B \to \left(\tan\left(\frac{\pi}{2}r\right)\cos\theta, \tan\left(\frac{\pi}{2}r\right)\sin\theta\right) \in \mathbb{R}^2 \quad \blacksquare \tag{2.52}$$

2.3.3 Topological space theory

When two spaces are homeomorphic, the mapping between them preserves the spaces' topological properties, such as set openness, closeness, and closure. This means that the topological properties are conserved as far as the convergence of point sequences is conserved under the mapping. This allows us to establish the topology of a space as the starting point of all topics presented in this chapter. We can base our study of topology on the topological properties of the set rather than the distance between points. This idea has led to one of the major fields of mathematics originating in the early twentieth century, called **topological space theory**.

Using topological properties as a starting point may be uncomfortable for those of us familiar with the concept of distance. Therefore, we will briefly explore this concept but will not use it as the basis of our study.

Definition (Topological space):
Consider a collection Θ of subsets[17] of a given set X, and denote its elements by

$$O_r \in \Theta (r = 1, 2, \ldots). \tag{2.53}$$

Suppose Θ and O_r satisfy:
1. $O_1 \cup O_2 \cup O_3 \cup \cdots \in \Theta$, 2. $O_1, O_2 \in \Theta \Rightarrow O_1 \cap O_2 \in \Theta$ 3. $X \in \Theta$ 4. $\varnothing \in \Theta$.
Then X is called a topological space[18]. ■

Most importantly, X should be a very abstract set, unrelated to the concept of distance. We can see there is no mention of distance in our definition. The definition of a topological space[19] (X, Θ) provides a collection of subsets that satisfies specific conditions. In topological space theory, this collection of subsets is the source of other topological concepts, such as closure, neighborhood, and continuous mapping, in which the concept of distance does not appear at all. However, all the consequences obtained through this theory's construction can be applied to our familiar spaces, such as three-dimensional Euclidean space and the one-dimensional number line.

---

[17] In mathematics, a collection of subsets of a given set is called a **family of sets** over X.

[18] In this case, an element of the family Θ is called an open set in X.

[19] The notation (X, Θ) refers to the topological space X whose topological properties are characterized by the family of sets Θ.

Chapter 3

Vector Space

3.1 What is vector space?

3.1.1 Beyond arrow-based representation

A **vector space** is a set of mathematical entities to which addition and multiplication can apply. For instance, the number sets $\mathbb{R}$ and $\mathbb{C}$ form vector spaces, because they allow calculations like "$6+2=8$" and "$(1+i)\times(2-i)=3+i$". Other mathematical entities that form a vector space include **functions**, **polynomials**, **linear operators**, as well as geometric arrows (=those which are commonly called "vectors" in elementary geometry).

Importantly, many theorems and formulae deduced from the vector space theory are universal in the sense that they hold true no matter what kinds of entities are chosen for the elements of the space. This universality plays a key role in the functional analysis, because all the appropriate entities are regarded as elements in vector spaces without conceptual distinction. Therefore, we ought to master the unified viewpoint on general vector spaces for which the class of constituent mathematical entities remain unspecified.

Keypoint: It's not the case that only geometric arrows form a vector space.

The nature of a vector space becomes enriched if several fundamental concepts are introduced to it. These include: distance between elements, norm, inner product, orthogonality, and completeness of a space. As a vector space gets endowed with those properties one-by-one, the space is going to demonstrate fructuous algebraic and geometric structures. The final goal of this flourishing procedure is to build **Hilbert spaces**, which is one of the pivotal vector space in the study of functional analysis. The salient feature of Hilbert spaces is that they are endowed with **inner product** and **completeness**. The concept of inner product provides the way how we should quantify similarities and differences between two elements contained in the same space. In addition, the concept of completeness open a gate for establishing a wide variety of **complete orthonormal sets**, using which we can decompose a given intractable mathematical entity into a set of easily-handled alternatives.[1] As a result, a large number of problems in physics and engineering can be treated with a geometric perspective in Hilbert spaces.

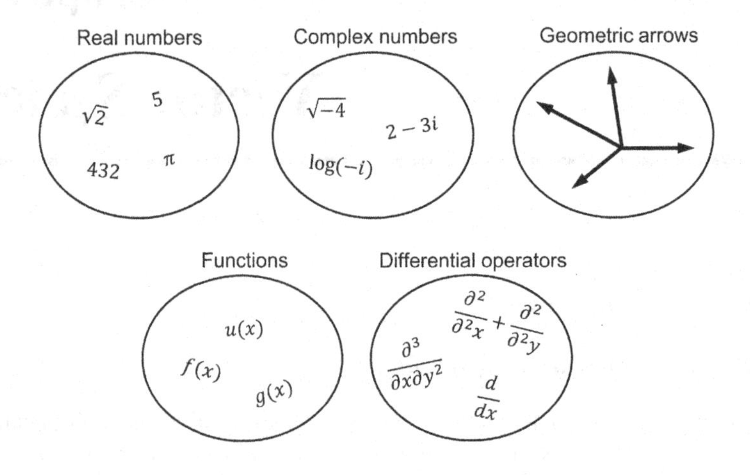

Figure 3.1: Vector spaces are formed by many kinds of mathematical entities. Five examples of vector spaces are shown here.

In the remainder of this chapter, we will survey precise definition of vector spaces and other related materials, followed by several important theorems that are relevant to understand the nature of Hilbert spaces.

3.1.2 Axiom of vector spaces

To make the contents self-contained, we first give a precise description of vector spaces, though it might seem lengthy at first glance. Emphasis is laid on the fact that the term "vector space" refers not exclusively to a set of geometric *arrows* which appear frequently in elementary linear algebra. Instead, it refers to general mathematical systems whose structures are regulated by specific algebraic rules demonstrated below.

[1]The primary example of such decomposition technique is the **Fourier series expansion**, wherein trigonometric functions serve as a complete orthonormal sets. The Fourier series expansion reduces many mathematical problems into more simpler ones, and its efficiency is manifested particularly in solving differential equations.

Definition (Vector space):
A **vector space** V is a collection of elements $\boldsymbol{x}$ (called **vectors**) that satisfy the following axioms:

1. V is a commutative[2] group[3] under addition:
- i) $\boldsymbol{x}+\boldsymbol{y}=\boldsymbol{y}+\boldsymbol{x}\in V$ for any $\boldsymbol{x},\boldsymbol{y}\in V$. [Closure]
- ii) $\boldsymbol{x}+(\boldsymbol{y}+\boldsymbol{z})=(\boldsymbol{x}+\boldsymbol{y})+\boldsymbol{z}\in V$ for any $\boldsymbol{x},\boldsymbol{y},\boldsymbol{z}\in V$. [Associativity]
- iii) The zero vector $\mathbf{0}$ exists for every $\boldsymbol{x}\in V$ such that $\boldsymbol{x}+\mathbf{0}=\boldsymbol{x}$. [Identity]
- iv) Additive inverse $-\boldsymbol{x}$ exists for every $\boldsymbol{x}\in V$ such that $\boldsymbol{x}+(-\boldsymbol{x})=\mathbf{0}$. [Invertibility]

2. V satisfies the following axioms with respect to a field[4] $\mathbb{F}$, whose elements are called **scalars**:
- i) V is closed under scalar multiplication:

$$\alpha\boldsymbol{x}\in V \quad \text{for arbitrary } \boldsymbol{x}\in V \text{ and } \alpha\in\mathbb{F}. \tag{3.1}$$

- ii) Scalar multiplication is **distributive** with respect to elements of both V and $\mathbb{F}$:

$$\alpha(\boldsymbol{x}+\boldsymbol{y})=\alpha\boldsymbol{x}+\alpha\boldsymbol{y}\in V, \quad (\alpha+\beta)\boldsymbol{x}=\alpha\boldsymbol{x}+\beta\boldsymbol{x}\in V. \tag{3.2}$$

- iii) Scalar multiplication is **associative**: $\alpha(\beta\boldsymbol{x})=\beta(\alpha\boldsymbol{x})$.
- iv) Multiplication with the zero scalar $0\in\mathbb{F}$ gives the zero vector such that $0\boldsymbol{x}=\mathbf{0}\in V$.
- v) The unit scalar $1\in\mathbb{F}$ has the property that $1\boldsymbol{x}=\boldsymbol{x}$. ■

In the definition above, the field $\mathbb{F}$ is either the set of real numbers $\mathbb{R}$ or the set of complex numbers $\mathbb{C}$. A vector space over $\mathbb{R}$ is called a **real vector space**. If $\mathbb{F}=\mathbb{C}$, then V is called a **complex vector space**.

Keypoint:
1. A vector space is a "group" under addition.
2. Its axiom relies on the algebraic concept of "field".

---

[2] Two elements are said to be **commutative** under a given operation if changing the order of two operands does not change the result. For instance, two complex numbers u and v are commutative under addition and multiplications, because $u+v=v+u$ and $uv=vu$.

[3] In mathematics, a **group** is a set of elements together with an operation that combines any two of elements to form a third element satisfying four conditions: closure, associativity, identity, and invertibility.

[4] Roughly speaking, a **field** is an algebraic structure with notions of addition, subtraction, multiplication, and division.

3.1.3 *Example of vector spaces*

A few simple examples of vector spaces are given below.

Example 1:
The set of all real numbers $\mathbb{R}$ forms a vector space over $\mathbb{R}$. This is proved by the fact that for any $x_1, x_2 \in \mathbb{R}$ and any $\alpha \in \mathbb{R}$, $x_1 + x_2$ and αx_1 are real numbers. ■

Example 2:
The set of all complex numbers $\mathbb{C}$ forms a vector space over $\mathbb{C}$. Indeed for any $z_1, z_2 \in \mathbb{C}$ and any $\alpha \in \mathbb{C}$, we have $z_1 + z_2 \in \mathbb{C}$ and $\alpha z_1 \in \mathbb{C}$. ■

Example 3:
The set of all n-tuples of complex numbers denoted by

$$\boldsymbol{x} = (\xi_1, \xi_2, \cdots, \xi_n) \tag{3.3}$$

forms a vector space if addition of vectors and multiplication of a vector by a scalar are defined by

$$\begin{aligned} \boldsymbol{x} + \boldsymbol{y} &= (\xi_1, \xi_2, \cdots, \xi_n) + (\eta_1, \eta_2, \cdots, \eta_n) \\ &= (\xi_1 + \eta_1, \xi_2 + \eta_2, \cdots, \xi_n + \eta_n), \qquad (3.4) \\ \alpha \boldsymbol{x} &= \alpha(\xi_1, \xi_2, \cdots, \xi_n) = (\alpha\xi_1, \alpha\xi_2, \cdots, \alpha\xi_n). \quad ■ \qquad (3.5) \end{aligned}$$

Example 4:
Consider the set of all polynomials $\{p_i(x)\}$ in a real variable x, whose elements are represented by

$$p_i(x) = c_0^{(i)} + c_1^{(i)} x + c_2^{(i)} x^2 + \cdots + c_{m_i}^{(i)} x^{m_i}, \quad c_j^{(i)} \in \mathbb{C}. \tag{3.6}$$

Here, the values of $c_j^{(i)}$ and m_i depends on i and/or j. Then the set of the polynomials forms a complex vector space. For instance, addition of the two polynomials

$$p_1(x) = 1 + 2ix + x^3, \quad p_2(x) = 3x - 4ix^2, \tag{3.7}$$

results in another polynomial given by

$$p_3(x) = 1 + (3 + 2i)x - 4ix^2 + x^3. \quad ■ \tag{3.8}$$

3.2 Property of vector space

3.2.1 *Inner product*

Remember that the axioms of vector spaces give no information as to the way of measuring the length of a vector and the relative angle between two vectors. To obtain geometric understanding, we need criterion for the length and angle of an element even in abstract vector spaces. This need is fulfilled by introducing the concept of **inner product**, which enriches the structure of a vector space enormously as we will see later.

Definition (Inner product):
An inner product is a mapping from an ordered pair of vectors $\boldsymbol{x}$ and $\boldsymbol{y}$, written by $(\boldsymbol{x},\boldsymbol{y})$, to a scalar (complex-valued, in general). The mapping satisfies the following rules.[5]

1. $(\boldsymbol{x},\boldsymbol{y}) = (\boldsymbol{y},\boldsymbol{x})^*$;
2. $(\alpha\boldsymbol{x}+\beta\boldsymbol{y},\boldsymbol{z}) = \alpha^*(\boldsymbol{x},\boldsymbol{z})+\beta^*(\boldsymbol{y},\boldsymbol{z})$, where α and β are certain complex numbers;
3. $(\boldsymbol{x},\boldsymbol{x}) \geq 0$ for any $\boldsymbol{x}$;
4. $(\boldsymbol{x},\boldsymbol{x}) = 0$ if and only if $\boldsymbol{x} = 0$. ■

Definition (Inner product space):
Vector spaces endowed with an inner product are called **inner product spaces.**[6] ■

In the definition above, the symbol (,) was used to represent a mapping from paired vectors to a complex number. But how can we evaluate the complex number to be mapped? Importantly, there is no common way of evaluating the value; it depends on our choices and the nature of the constituent entities that form the vector space. This methodological versatility for the inner product evaluations makes it possible to create many useful function spaces (e.g., ℓ^2 space, L^2 space, wavelet space, Sobolev space,$\cdots$), each of which is endowed with different ways of inner product evaluations. Three simple examples of inner product spaces are given below.

Example 1:
The set of complex numbers $\{z_1,z_2,\cdots,z_n\}$, denoted by $\mathbb{C}^n$, is an inner product space. For two vectors $\boldsymbol{x} = (\xi_1,\xi_2,\cdots\xi_n)$ and $\boldsymbol{y} = (\eta_1,\eta_2,\cdots\eta_n)$, the inner product is defined by

$$(\boldsymbol{x},\boldsymbol{y}) = \sum_{i=1}^{n} \xi_i^* \eta_i. \quad ■ \tag{3.9}$$

Example 2:
The set of complex-valued continuous functions with respect to the real variable x on the closed interval $[a,b]$ is an inner product space. In this situation, the inner product between two functions $f(x)$ and $g(x)$ is defined by

$$(f,g) = \int_a^b f^*(x)g(x)dx. \quad ■ \tag{3.10}$$

[5] Here, the asterisk $(*)$ indicates to take **complex conjugate**.

[6] In particular, a real inner product space is called a **Euclidean space**, and a complex inner product space is called a **unitary space**.

Example 3:
The set of **Hermite polynomials**[7] $\{H_n(x)\}$ constitute an inner product space under the inner product defined by

$$(H_n, H_m) = \int_{-\infty}^{\infty} H_n(x)H_m(x)w(x)dx \ [= \sqrt{\pi}2^n n! \delta_{m,n}]. \tag{3.12}$$

Here, $\delta_{m,n}$ is **Kronecker's delta**[8] and $w(x) = e^{-x^2}$ is called a **weight function**.[9] ■

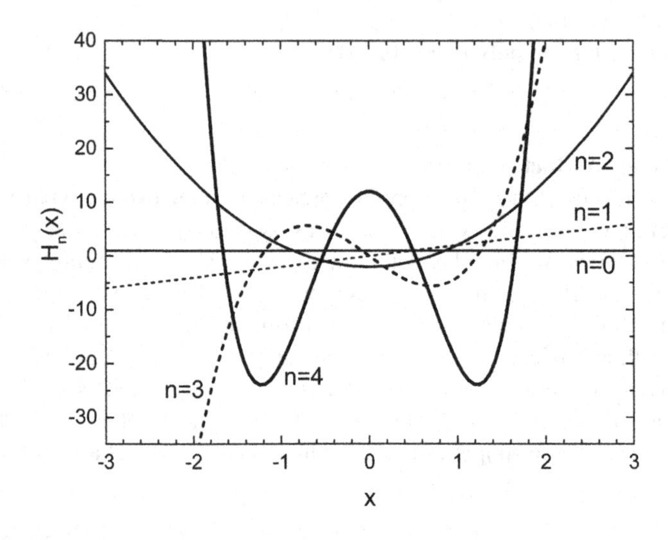

Figure 3.2: Profiles of the first five Hermite polynomials $H_n(x)$.

It is pedagogical to show a counter-example of inner product spaces. Consider a set of all column **four-vectors**. Its elements are expressed by

$$\boldsymbol{x} = [\xi_1, \xi_2, \xi_3, \xi_4] \quad \text{or} \quad \boldsymbol{y} = [\eta_1, \eta_2, \eta_3, \eta_4], \tag{3.14}$$

---

[7]A **Hermite polynomial** of the nth order is given by

$$H_n(x) = (-1)^n e^{x^2} \frac{d^n}{dx^n}(e^{-x^2}). \tag{3.11}$$

The first few terms are expressed by $H_0(x) = 1$, $H_1(x) = 2x$, $H_2(x) = 4x^2 - 2$, $H_3(x) = 8x^3 - 12x$, $H_4(x) = 16x^4 - 48x^2 + 12$.

[8]**Kronecker's delta** $\delta_{m,n}$ is defined by

$$\delta_{m,n} = \begin{cases} 1 & \text{if } m = n, \\ 0 & \text{otherwise.} \end{cases} \tag{3.13}$$

[9]The explicit form of a weight function depends on orthogonal polynomials: each of Jacobi-, Laguerre-, Chebyshev-, and many other orthogonal polynomials shows unique functional form of $w(x)$.

where ξ_i and η_i $(1 \le i \le 4)$ are real numbers. Sometimes we use the quantity

$$(\boldsymbol{x}, \boldsymbol{y}) \equiv \xi_1\eta_1 + \xi_2\eta_2 + \xi_3\eta_3 - \xi_4\eta_4, \tag{3.15}$$

which looks like the inner product of two vectors defined by (3.9). However, this is not the case because the minus sign is attached to the last term in the right hand side of (3.15). Hence, this quantity does not satisfy a portion of requirements for inner product. In fact, $(\boldsymbol{x}, \boldsymbol{x})$ under the definition of (3.15) can be zero even though $\boldsymbol{x} \neq \boldsymbol{0}$, contrary to the definition of inner product. Thus the quantity (3.15) is not an inner product; instead, it is called a **quasi-inner product** and is known to play an important role in the theory of special relativity.

Keypoint: Any inner product must possess positive definite property: $(\boldsymbol{x}, \boldsymbol{x}) > 0$ for $\boldsymbol{x} \neq \boldsymbol{0}$.

Another important feature of inner products is that they are not linear but **conjugate linear** with respect to the first factor; that is,

$$(\alpha\boldsymbol{x}, \boldsymbol{y}) = \alpha^*(\boldsymbol{x}, \boldsymbol{y}). \tag{3.16}$$

Furthermore, inner products are not symmetric in the sense that

$$(\boldsymbol{x}, \boldsymbol{y}) \neq (\boldsymbol{y}, \boldsymbol{x}) \tag{3.17}$$

but rather

$$(\boldsymbol{x}, \boldsymbol{y}) = (\boldsymbol{y}, \boldsymbol{x})^*. \tag{3.18}$$

The latter fact implies that $(\boldsymbol{x}, \boldsymbol{x})$ is real for every $\boldsymbol{x}$, and hence, we can define the *length* of the vector $\boldsymbol{x}$ by

$$||\boldsymbol{x}|| = \sqrt{(\boldsymbol{x}, \boldsymbol{x})}. \tag{3.19}$$

Since $(\boldsymbol{x}, \boldsymbol{x}) \ge 0$, $||\boldsymbol{x}||$ is always non-negative and real. The quantity $||\boldsymbol{x}||$ is referred to as the **norm** induced by the inner product $(\boldsymbol{x}, \boldsymbol{y})$. It also follows that

$$||\alpha\boldsymbol{x}|| = \sqrt{(\alpha\boldsymbol{x}, \alpha\boldsymbol{x})} = \sqrt{\alpha^*\alpha(\boldsymbol{x}, \boldsymbol{x})} = |\alpha| \cdot ||\boldsymbol{x}||. \tag{3.20}$$

Hence the norm of $\alpha\boldsymbol{x}$ equals to the norm of $\boldsymbol{x}$ multiplied by $|\alpha|$. This result agrees with our intuitive understanding of the length of an geometric arrow; the length of an arrow $\alpha\boldsymbol{x}$ equals to that of $\boldsymbol{x}$ multiplied by $|\alpha|$.

Precisely speaking, the quantity $\|\boldsymbol{x}\|$ introduced by (3.19) is only a special kind of norm, because it is induced by an inner product. In view of functional analysis, the norm should be more general concept and can be introduced independently of the concept of inner product. Two famous examples of norms with no reference to any inner product are ℓ^p norm and L^p norm, which will be defined later by (3.53) and (3.55), respectively. See sections 3.3.3 and 3.3.4 for details.

Keypoint: A norm is a broader concept than an inner product.

3.2.2 Geometry of inner product spaces

Once endowed with a norm and an inner product, vector spaces take on geometric structures based on which we can argue the length of an element and convergence of an infinite sequence of elements in the spaces. The structures obtained can be interpreted in view of analogy with elementary geometry, and characterized by the following three formulae. Meanwhile, we assume $\boldsymbol{x} \neq 0$ and $\boldsymbol{y} \neq 0$; otherwise the formulae all become trivial.

Theorem:
Given an inner product space, any two elements $\boldsymbol{x}$ and $\boldsymbol{y}$ satisfy the following equality/inequality.

$$|(\boldsymbol{x},\boldsymbol{y})| \leq \|\boldsymbol{x}\|\,\|\boldsymbol{y}\|; \qquad \textbf{[Schwarz's inequality]} \quad (3.21)$$

$$\|\boldsymbol{x}+\boldsymbol{y}\| \leq \|\boldsymbol{x}\| + \|\boldsymbol{y}\|; \qquad \textbf{[Triangle inequality]} \quad (3.22)$$

$$\|\boldsymbol{x}+\boldsymbol{y}\|^2 + \|\boldsymbol{x}-\boldsymbol{y}\|^2 = 2\left(\|\boldsymbol{x}\|^2 + \|\boldsymbol{y}\|^2\right). \quad \blacksquare \qquad \textbf{[Parallelogram law]} \quad (3.23)$$

Among the formulae above-mentioned, we first mention the intuitive explanation and practical merits of Schwarz's inequality.[10] Since $\|\boldsymbol{x}\| > 0$ under our assumption of $\boldsymbol{x} \neq \boldsymbol{0}$, the inequality is rewritten by

$$\|\boldsymbol{y}\| \geq \frac{|(\boldsymbol{x},\boldsymbol{y})|}{\|\boldsymbol{x}\|}. \tag{3.24}$$

In view of elementary geometry, the quantity $|(\boldsymbol{x},\boldsymbol{y})|/\|\boldsymbol{x}\|$ is the length of the projection of the arrow $\boldsymbol{y}$ onto $\boldsymbol{x}$ (See Fig. 3.3). Therefore, Schwarz's inequality tells us that this length cannot be greater than $\|\boldsymbol{y}\|$. Plainly speaking, it means that the hypotenuse in a right-angled triangle is not shorter than one of the other sides. It also follows from Fig. 3.3 that the equality in (3.24) holds only when $\boldsymbol{y}$ is parallel to $\boldsymbol{x}$. This is true in abstract vector spaces, too. In the latter cases, the phrase "$\boldsymbol{y}$ is parallel to $\boldsymbol{x}$" should be replaced by "$\boldsymbol{y}$ is linearly dependent on $\boldsymbol{x}$", as represented by $\boldsymbol{y} = \beta\boldsymbol{x}$ wherein β is a complex-valued constant.

Schwarz's inequality is crucially important in the study of functional analysis, because it is often used when discussing convergence of an infinite sequence of elements in inner product spaces. Also, practical importance of Schwarz's inequality arises in solving differential equations. For instance, if f and g are square-integrable functions on $\mathbb{R}$, then Schwarz's inequality says

$$\left|\int_{-\infty}^{\infty} f(x)g(x)dx\right|^2 \leq \int_{-\infty}^{\infty} |f(x)|^2 dx \cdot \int_{-\infty}^{\infty} |g(x)|^2 dx. \tag{3.25}$$

This type of inequality is frequently involved in the proof of the existence of solutions of a given differential equation. Furthermore, an important physical application

[10] It is also known as the **Cauchy-Schwarz inequality**.

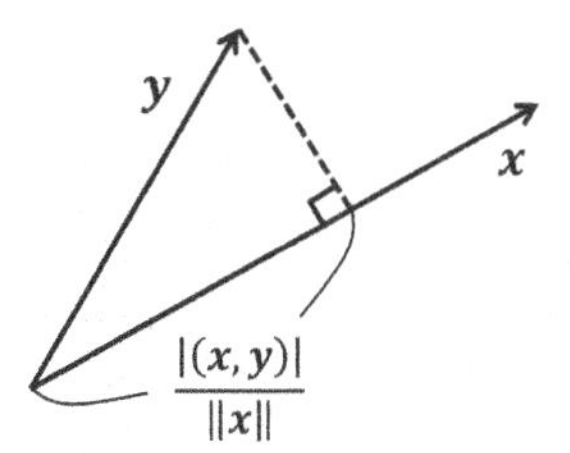

Figure 3.3: Schematic illustration of Schwarz's inequality in the sense of elementary geometry.

of Schwarz's inequality is found in the general formulation of the **uncertainty principle.**[11] The principle is a consequence of a fundamental postulate of quantum mechanics that all quantum phenomena must be described in terms of wave-particle duality.

Proof (of Schwarz's inequality):

From the definition of the inner product, we see that for any α,

$$0 \leq (\boldsymbol{x}+\alpha\boldsymbol{y}, \boldsymbol{x}+\alpha\boldsymbol{y}) = (\boldsymbol{x},\boldsymbol{x}) + \alpha(\boldsymbol{x},\boldsymbol{y}) + \alpha^*(\boldsymbol{y},\boldsymbol{x}) + |\alpha|^2(\boldsymbol{y},\boldsymbol{y}). \tag{3.26}$$

Put $\alpha = -(\boldsymbol{x},\boldsymbol{y})/(\boldsymbol{y},\boldsymbol{y})$ and multiply by $(\boldsymbol{y},\boldsymbol{y})$ to obtain

$$0 \leq (\boldsymbol{x},\boldsymbol{x})(\boldsymbol{y},\boldsymbol{y}) - |(\boldsymbol{x},\boldsymbol{y})|^2. \tag{3.27}$$

This gives Schwarz's inequality of (3.21). The equality holds if and only if $\boldsymbol{x}$ and $\boldsymbol{y}$ are linearly dependent.[12] ■

---

[11] A few of the more common pairs in the uncertainty principle are:

i)Between the position and momentum of an object. The product of the standard deviation of the ith component of the position $\Delta\hat{x}_i$ and that of momentum $\Delta\hat{p}_i$ satisfies the inequality $\Delta\hat{x}_i\Delta\hat{p}_i \geq \hbar/2$, where $\hbar$ is a physical constant called **Plank's constant**.

ii)Between two orthogonal components of the total angular momentum operator of an object: $\Delta\hat{J}_i\Delta\hat{J}_j \geq (\hbar/2)\left|\langle\hat{J}_k\rangle\right|$, where $\hat{J}_i$ denotes angular momentum along the x_i axis and angular brackets indicate to take an expectation value.

iii)Between the number of electrons (represented by $\hat{N}$) in a superconductor and the phase of its order parameter (by $\hat{\phi}$): $\Delta\hat{N}\Delta\hat{\phi} \geq 1$.

[12] If $\boldsymbol{x}$ and $\boldsymbol{y}$ are linearly dependent, then $\boldsymbol{y} = \beta\boldsymbol{x}$ for a complex number β so that we have

$$|(\boldsymbol{x},\boldsymbol{y})| = |(\boldsymbol{x},\beta\boldsymbol{x})| = |\beta|(\boldsymbol{x},\boldsymbol{x}) = |\beta|\|\boldsymbol{x}\|\,\|\boldsymbol{x}\| = \|\boldsymbol{x}\|\,\|\beta\boldsymbol{x}\| = \|\boldsymbol{x}\|\,\|\boldsymbol{y}\|. \tag{3.28}$$

The converse is also true; if $|(\boldsymbol{x},\boldsymbol{y})| = \|\boldsymbol{x}\|\,\|\boldsymbol{y}\|$, we have

$$|(\boldsymbol{x},\boldsymbol{y})|^2 = (\boldsymbol{x},\boldsymbol{y})(\boldsymbol{y},\boldsymbol{x}) = (\boldsymbol{x},\boldsymbol{x})(\boldsymbol{y},\boldsymbol{y}) = \|\boldsymbol{x}\|^2\,\|\boldsymbol{y}\|^2. \tag{3.29}$$

This implies that

$$\|(\boldsymbol{y},\boldsymbol{y})\boldsymbol{x} - (\boldsymbol{y},\boldsymbol{x})\boldsymbol{y}\|^2 = \|\boldsymbol{y}\|^4\|\boldsymbol{x}\|^2 + |(\boldsymbol{y},\boldsymbol{x})|^2\,\|\boldsymbol{y}\|^2 - \|\boldsymbol{y}\|^2(\boldsymbol{y},\boldsymbol{x})(\boldsymbol{x},\boldsymbol{y}) - \|\boldsymbol{y}\|^2(\boldsymbol{y},\boldsymbol{x})^*(\boldsymbol{y},\boldsymbol{x}) = 0, \tag{3.30}$$

where (3.29) and the relation $(\boldsymbol{y},\boldsymbol{x})^* = (\boldsymbol{x},\boldsymbol{y})$ were used. The result (3.30) means $(\boldsymbol{y},\boldsymbol{y})\boldsymbol{x} = (\boldsymbol{x},\boldsymbol{y})\boldsymbol{y}$, which clearly shows that $\boldsymbol{x}$ and $\boldsymbol{y}$ are linearly dependent.

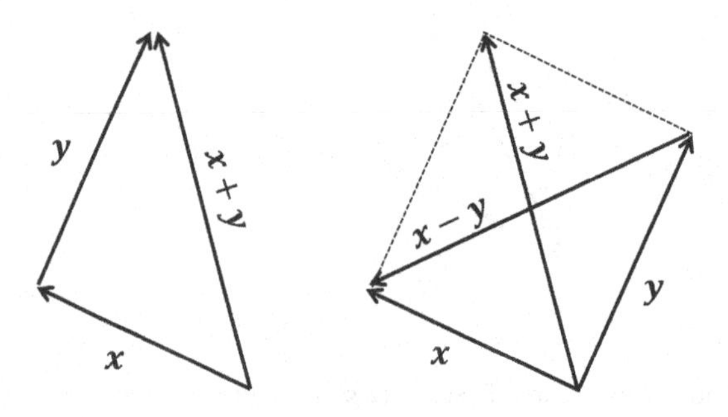

Figure 3.4: Geometric arrow configuration illustrating the triangle's inequality and the parallelogram law.

It should be emphasized that the proof above-mentioned is based on only the axioms of inner product. Accordingly, Schwarz's inequality holds true regardless of the way how we evaluate the value of inner products.

Next we have an intuitive understanding of the triangle inequality. In a sense of elementary geometry, it means that: "If you want to go from one point to another, then the shortest path is the straight line between the two points." This holds true even in abstract inner product spaces. Using the triangle inequality, we are able to find the shortest path between a pair of elements in abstract vector spaces. Proof of the inequality is given below.

Proof (of Triangle inequality):

Putting $\alpha = 1$ in (3.26), we have

$$\begin{aligned} \|\boldsymbol{x}+\boldsymbol{y}\|^2 &= (\boldsymbol{x},\boldsymbol{x})+(\boldsymbol{y},\boldsymbol{y})+2\mathrm{Re}(\boldsymbol{x},\boldsymbol{y}) \le (\boldsymbol{x},\boldsymbol{x})+(\boldsymbol{y},\boldsymbol{y})+2|(\boldsymbol{x},\boldsymbol{y})| \\ &\le \|\boldsymbol{x}\|^2+\|\boldsymbol{y}\|^2+2\|\boldsymbol{x}\|\|\boldsymbol{y}\| \quad \text{(by Schwarz's inequality)} \\ &= (\|\boldsymbol{x}\|+\|\boldsymbol{y}\|)^2. \end{aligned}$$

This proves the desired inequality of (3.22). ■

Finally we comment on an intuitive interpretation of the **parallelogram law**.[13] Again in the elementary sense, it states that, given a parallelogram, the sum of the squared lengths of its four sides $(= 2\|\boldsymbol{x}\|^2 + 2\|\boldsymbol{y}\|^2)$ equals the sum of the squared lengths of its two diagonals $(= \|\boldsymbol{x}+\boldsymbol{y}\|^2 + \|\boldsymbol{x}-\boldsymbol{y}\|^2)$. Particularly when the parallelogram is a rectangle, the statement reduces to the **Pythagorean theorem**.

Proof (of Parallelogram law):

Straightforward expansion yields

$$\|\boldsymbol{x}+\boldsymbol{y}\|^2 = \|\boldsymbol{x}\|^2 + (\boldsymbol{x},\boldsymbol{y}) + (\boldsymbol{y},\boldsymbol{x}) + \|\boldsymbol{y}\|^2. \tag{3.31}$$

[13] Also called the **parallelogram identity**. Its elementary version is known as **Pappus's theorem**.

Replace $\boldsymbol{y}$ by $-\boldsymbol{y}$ to obtain

$$\|\boldsymbol{x}-\boldsymbol{y}\|^2 = \|\boldsymbol{x}\|^2 - (\boldsymbol{x},\boldsymbol{y}) - (\boldsymbol{y},\boldsymbol{x}) + \|\boldsymbol{y}\|^2. \tag{3.32}$$

By adding (3.31) and (3.32), we attain our objective of (3.23). ■

Keypoint: All the elements in inner product spaces satisfy: Schwarz's inequality, the triangle inequality, and the parallelogram law.

3.2.3 *Orthogonality and orthonormality*

One of the most important consequence of having the inner product is to be able to define the **orthogonality** of vectors. In linear algebra, orthogonality simply means two arrows are perpendicular to each other. In functional analysis, on the other hand, the "relative angle" between two elements cannot be defined unless an appropriate rule is imposed. The need is fulfilled by the introduction of inner product, as stated below.

Definition (Orthogonality):
Two elements $\boldsymbol{x}$ and $\boldsymbol{y}$ in an inner product space are said to be **orthogonal** if and only if $(\boldsymbol{x},\boldsymbol{y}) = 0$. ■

Example:
The two functions $f(x) = \sin x$ and $g(x) = \cos x$ defined on the closed interval $[-\pi,\pi]$ are orthogonal to each other in the sense that

$$(f,g) = \int_{-\pi}^{\pi} \sin x \cos x dx = 0. \quad ■ \tag{3.33}$$

Figure 3.5 illustrates the reason of the orthogonality between $f(x) = \sin x$ and $g(x) = \cos x$. It shows that the product of the two functions, $f(x)g(x) = \sin x \cos x$ oscillates around the horizontal line of $y = 0$, wherein the deviation above the line is canceled out by the deviation below the line. Hence the oscillation gives no contribution to the integral, as a result of which the integral of $\int_{-\pi}^{\pi} \sin x \cos x dx$ equals to zero.

Notably, if

$$(\boldsymbol{x},\boldsymbol{y}) = 0, \tag{3.34}$$

then

$$(\boldsymbol{x},\boldsymbol{y}) = (\boldsymbol{y},\boldsymbol{x})^* = 0 \tag{3.35}$$

so that $(\boldsymbol{y},\boldsymbol{x}) = 0$ as well. Thus, the orthogonality is a symmetric relation, although the inner product is asymmetric. Note also that the **zero vector 0** is orthogonal to every vector in the inner product space.

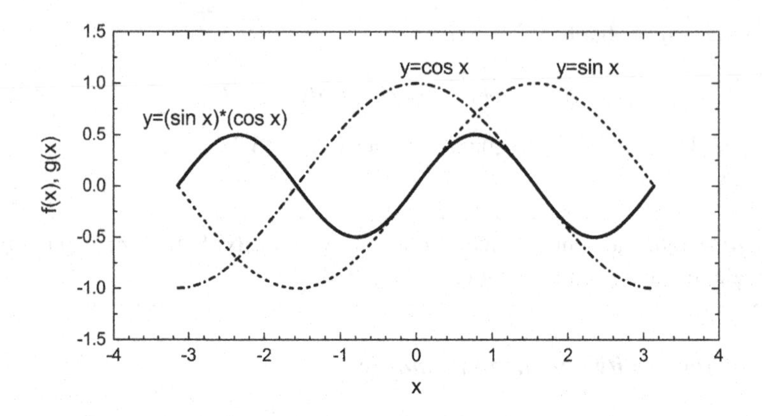

Figure 3.5: Orthogonal relation between $f(x) = \sin x$ and $g(x) = \cos x$.

Realize that any vector $\boldsymbol{x}$ can be normalized by dividing it by its length to form the new vector $\boldsymbol{x}/||\boldsymbol{x}||$ with unit length. Combination of the two properties, orthogonality and normalization, leads us to the concept of **orthonormality**.

Definition (Orthonormal set of vectors):
A set of n vectors $\{\boldsymbol{x}_1, \boldsymbol{x}_2, \cdots \boldsymbol{x}_n\}$ is called **orthonormal** if

$$(\boldsymbol{x}_i, \boldsymbol{x}_j) = \delta_{ij} \tag{3.36}$$

for all i and j. ■

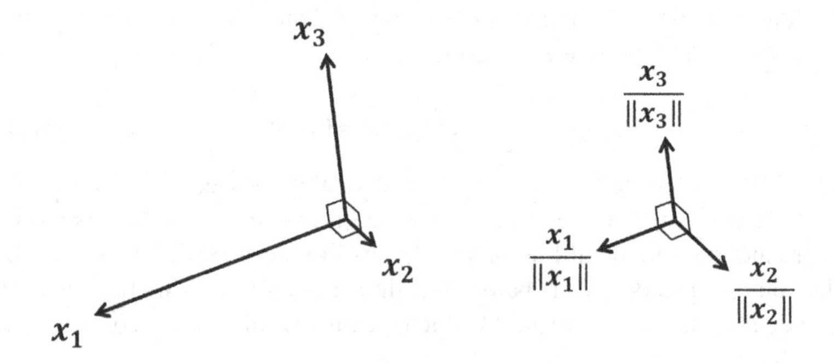

Figure 3.6: Left: Orthogonal set of vectors. Right: Ortho"normal" set of vectors.

That is, the orthonormality of a set of vectors means that each vector is orthogonal to all others in the set and is normalized to unit length. An example of an orthonormal set of vectors is the set of three unit vectors, $\{e_i\}$ $(i = 1,2,3)$, for the three-dimensional **Cartesian space**. The Cartesian space is a Euclidean space with rectangular coordinates, using which a point in the space is referred to. The coordinates are three real numbers, indicating the positions of the perpendicular projections from the point to three mutually-perpendicular axes. Another set of orthonormal set is a set of functions $\{e^{in\theta}/\sqrt{2\pi}\}$ with respect to θ defined in the closed interval $[-\pi,\pi]$. The inner product reads

$$\int_{-\pi}^{\pi} \frac{e^{-im\theta}}{\sqrt{2\pi}} \cdot \frac{e^{in\theta}}{\sqrt{2\pi}} d\theta = \begin{cases} 1 & \text{if } m=n, \\ 0 & \text{otherwise.} \end{cases} \tag{3.37}$$

indicating that all the elements are mutually orthogonal.

3.2.4 Linear independence

Meanwhile, we consider the case of finite dimension, i.e., the set consisting of finite number elements. Preliminary study on such finite-dimensional space will make clear the fundamental difference in the nature of inner produce spaces with finite and infinite dimension.

Definition (Linear independence in finite dimension):
A *finite*[14] set of vectors, say, $\boldsymbol{e}_1, \boldsymbol{e}_2, \cdots, \boldsymbol{e}_n$ is **linearly independent** if and only if

$$\sum_{i=1}^{n} c_i \boldsymbol{e}_i = 0 \;\Rightarrow\; c_i = 0 \text{ for all } i. \quad \blacksquare \tag{3.38}$$

When $\{\boldsymbol{e}_i\}$ is not linearly independent, there exists a set of coefficients $\{c_i\}$ that satisfies $\sum_{i=1}^{n} c_i \boldsymbol{e}_i = 0$ but $c_i \neq 0$ for certain *is*. This situation is realized by imposing $\boldsymbol{e}_2 = \beta \boldsymbol{e}_1$ with $\beta = -c_1/c_2 \neq 0$, for instance; then we have

$$\begin{aligned} & c_1\boldsymbol{e}_1 + c_2\boldsymbol{e}_2 + 0\boldsymbol{e}_3 + 0\boldsymbol{e}_4 + \cdots + 0\boldsymbol{e}_n \\ = \; & (c_1 + \beta c_2)\boldsymbol{e}_1 + 0\boldsymbol{e}_3 + 0\boldsymbol{e}_4 + \cdots + 0\boldsymbol{e}_n \\ = \; & 0\boldsymbol{e}_1 + 0\boldsymbol{e}_3 + 0\boldsymbol{e}_4 + \cdots + 0\boldsymbol{e}_n \\ = \; & \boldsymbol{0}, \end{aligned} \tag{3.39}$$

even though $c_1 \neq 0$ and $c_2 \neq 0$.

Theorem: An orthonormal set is linearly independent. $\blacksquare$

Proof:
Suppose that a set $\{\boldsymbol{x}_1, \boldsymbol{x}_2, \cdots, \boldsymbol{x}_n\}$ is orthonormal and satisfies the relation $\sum_{i=1}^{n} \alpha_i \boldsymbol{x}_i = \boldsymbol{0}$. Then, the orthonormal condition $(\boldsymbol{x}_i, \boldsymbol{x}_j) = \delta_{ij}$ results in that for any j,

$$0 = \left(\boldsymbol{x}_j, \sum_{i=1}^{n} \alpha_i \boldsymbol{x}_i\right) = \sum_{i=1}^{n} \alpha_i (\boldsymbol{x}_j, \boldsymbol{x}_i) = \sum_{i=1}^{n} \alpha_i \delta_{ij} = \alpha_j. \tag{3.40}$$

[14]The definition applies to *infinite* sets of vectors $\boldsymbol{e}_1, \boldsymbol{e}_2, \cdots$ if the vector space under consideration admits a definition of convergence; see section 4.1.1 for details.

Therefore, the set is linearly independent. ■

Importantly, the above theorem suggests that any orthonormal set serves as a **basis** of an inner product space.

Definition (Basis in finite dimension):
A **basis** of the vector space V is a set of linearly independent vectors $\{e_i\}$ of V such that every vector $\boldsymbol{x}$ of V can be expressed as

$$\boldsymbol{x} = \sum_{i=1}^{n} \alpha_i e_i. \tag{3.41}$$

Here, the numbers $\alpha_1, \alpha_2, \cdots, \alpha_n$ are **coordinates** of the vector $\boldsymbol{x}$ with respect to the basis, and they are uniquely determined due to the linear independence property. ■

Therefore, every set of n linearly independent vectors is a basis in an n-dimensional vector space. The number of n is called the **dimension** of the vector space. Obviously, an infinite-dimensional vector space does not admit a basis that consists of finite number vectors; this is why it is called infinite-dimensional.

Below is another consequence of the orthonormal set of vector.

Theorem (Bessel's inequality in finite dimension):
Suppose that $\{\boldsymbol{x}_1, \boldsymbol{x}_2, \cdots, \boldsymbol{x}_n\}$ is a set of orthonormal vectors in an inner product space. If $\boldsymbol{x}$ is any vector in the same space, then

$$||\boldsymbol{x}||^2 \geq |(\boldsymbol{x}_1, \boldsymbol{x})|^2 + |(\boldsymbol{x}_2, \boldsymbol{x})|^2 + \cdots + |(\boldsymbol{x}_n, \boldsymbol{x})|^2. \quad ■ \tag{3.42}$$

Proof:
Set $r_i = (\boldsymbol{x}_i, \boldsymbol{x})$ and[15] $\boldsymbol{x}' = \boldsymbol{x} - \sum_{i=1}^{n} r_i \boldsymbol{x}_i$ to obtain the inequality

$$\begin{aligned}
0 \leq ||\boldsymbol{x}'||^2 &= (\boldsymbol{x}', \boldsymbol{x}') = \left(\boldsymbol{x} - \sum_{i=1}^{n} r_i \boldsymbol{x}_i,\ \boldsymbol{x} - \sum_{j=1}^{n} r_j \boldsymbol{x}_j \right) \\
&= (\boldsymbol{x}, \boldsymbol{x}) - \sum_{i=1}^{n} r_i^* (\boldsymbol{x}_i, \boldsymbol{x}) - \sum_{j=1}^{n} r_j (\boldsymbol{x}, \boldsymbol{x}_j) + \sum_{i,j=1}^{n} r_i^* r_j (\boldsymbol{x}_i, \boldsymbol{x}_j) \\
&= ||\boldsymbol{x}||^2 - \sum_{i=1}^{n} |r_i|^2 - \sum_{j=1}^{n} |r_j|^2 + \sum_{j=1}^{n} |r_j|^2 \\
&= ||\boldsymbol{x}||^2 - \sum_{i=1}^{n} |r_i|^2.
\end{aligned}$$

This result is equivalent to (3.42). ■

It is also noteworthy that the equality between both sides of (3.42) holds only when the number n equals to the dimension of the space considered. Particularly

[15] The vector $\boldsymbol{x}'$ defined here is orthogonal to every $\boldsymbol{x}_j$, because for any $\boldsymbol{x}_j$ we have $(\boldsymbol{x}', \boldsymbol{x}_j) = (\boldsymbol{x}, \boldsymbol{x}_j) - \sum_{i=1}^{n} r_i^* (\boldsymbol{x}_i, \boldsymbol{x}_j) = r_j^* - r_j^* = 0$.

in this case, Bessel's inequality reduces to **Parseval's identity**, which is an abstract version of the well-known **Pythagorean theorem**. In other words, Bessel's inequality is true even when n is less than the dimension of the space. Specifically when we pick up only one element (e.g., $\boldsymbol{x}_1$), Bessel's inequality becomes Schwarz's inequality as shown below.

$$\|\boldsymbol{x}_1\| \cdot \|\boldsymbol{x}\| = \sqrt{\|\boldsymbol{x}\|^2} \geq \sqrt{|(\boldsymbol{x}_1, \boldsymbol{x})|^2} = |(\boldsymbol{x}_1, \boldsymbol{x})|. \tag{3.43}$$

In the derivation above, $\|\boldsymbol{x}_1\| = 1$ was used.

Keypoint: Bessel's inequality is reduced to **Parseval's identity**, if the number of summed terms (n) equals to the dimension of the space. It is further reduced to **Schwarz's inequality** if $n = 1$.

3.2.5 Complete vector spaces

Accompanied by the notion of inner product, the completeness is the salient features inherent to Hilbert spaces. When a vector space is in *finite* dimension, the completeness of the space can be proved by finding an orthonormal set that is not contained in another larger orthonormal set in the same space. This is intuitively understood by considering the three-dimensional Euclidean space; it suffices to find a set of three orthonormal vectors whose linear combinations represent any vector in the space. But when considering an *infinite*-dimensional space, it becomes nontrivial whether an infinite number of orthonormal vectors have no excess or deficiency. In fact, linear combinations of an infinite number of vectors sometimes deviate from the space considered, and at other times infinitely many vectors are not enough to represent a vector contained in the same space. These ambiguity can be removed by confirming the completeness of the infinite dimensional space to be considered.

In general, the completeness of a space is examined using the **Cauchy criterion** that is presented in Appendix B.2. Here is given a preparatory definition.

Definition (Cauchy sequence of vectors):
A sequence $\{\boldsymbol{x}_1, \boldsymbol{x}_2, \cdots\}$ of vectors is called a **Cauchy sequence** of vectors if for any positive $\varepsilon > 0$, there exists an appropriate number N such that $\|\boldsymbol{x}_m - \boldsymbol{x}_n\| < \varepsilon$ for all $m, n > N$. ■

In plain words, a sequence is a Cauchy sequence if the terms $\boldsymbol{x}_m$ and $\boldsymbol{x}_n$ in the sequence get closer unlimitedly as $m, n \to \infty$.

Definition (Convergence of an infinite vector sequence):
An infinite sequence of vectors $\{\boldsymbol{x}_1, \boldsymbol{x}_2, \cdots\}$ is said to be convergent if there exists an element $\boldsymbol{x}$ such that $\|\boldsymbol{x}_n - \boldsymbol{x}\| \to 0$. ■

Definition (Completeness of a vector space):
If every Cauchy sequence in a vector space is convergent, we say that the vector space is **complete**. ■

Since we currently focuses on inner product spaces, we have used the norm $\|x\| = \sqrt{(x,x)}$ induced by an inner product, in order to define the completeness. However, completeness apply more general vector spaces in which even an inner product is unnecessary. In fact, in the study of abstract vector spaces, we can define complete space that is free from inner product.

3.3 Hierarchy of vector space

3.3.1 From a mere collection of elements

In this section, we survey the hierarchical structure of abstract spaces. The hierarchy is established by step-by-step mounting of essential concepts like topology, distance, norm, inner product, and completeness to an initially mere collection of elements (= the most general and abstract space). As go deeper into the hierarchy, we will encounter various kinds of spaces, and Hilbert spaces are found at the deepest layer to be a very limited, special class of vector spaces under strict conditions.

Let us begin with a mere collection of elements (points, numbers, functions, operators, or other mathematical entities) for which neither topological nor algebraic rules apply yet. At this stage, we have no way of considering the relationship between elements, because there is no criterion to carry out. Once endowing the concept of topology with the set of elements, we become able to argue the inclusion relation of elements (described by open sets, closed sets, closures, neighborhoods,... etc.) and convergence of a sequence of elements; the resulting set of elements is called a **topological space**. Or, if you equip the concept of vector addition and scalar multiplication (*i.e.,* the vector space axiom), then the initially mere collection of elements mutates into a vector space.

Note that the two concepts, topology and axiom for vectors, are independent of each other (See Fig. 3.7). By introducing both the two concepts into the initially mere collection of elements, we obtain a **topological vector space.**[16] It should be emphasized that in topological vector spaces, we have no way of measuring the distance between elements. Neither topology nor vector axioms give no clue to evaluate the distance between two different elements; therefore, introducing the concept of distance is requisite for the quest to establishing mathematically important subspaces including Hilbert spaces.

[16]Or called **linear topological space**.

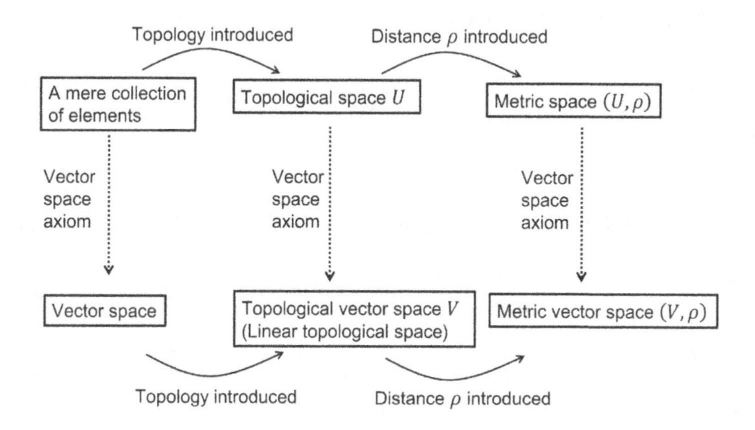

Figure 3.7: Operation sheet for constructing a metric vector space.

3.3.2 Metric vector spaces

To proceed our argument, let us introduce the concept of distance into topological vector spaces.

Definition (Distance function):
Suppose that two elements $\boldsymbol{x}, \boldsymbol{y}$ in a topological vector space V are mapped to a non-negative real number a through a function ρ defined by

$$\rho : \{\boldsymbol{x}, \boldsymbol{y}\} \to a \in \mathbb{R}. \tag{3.44}$$

This function is called a **distance function**[17] if it satisfies the conditions:

1. $\rho(\boldsymbol{x}, \boldsymbol{y}) = 0$ if and only if $\boldsymbol{x} = \boldsymbol{y}$,
2. $\rho(\boldsymbol{x}, \boldsymbol{y}) = \rho(\boldsymbol{y}, \boldsymbol{x})$, and
3. $\rho(\boldsymbol{x}, \boldsymbol{y}) \le \rho(\boldsymbol{x}, \boldsymbol{z}) + \rho(\boldsymbol{z}, \boldsymbol{y})$ for any $\boldsymbol{z} \in V$. ■

Definition (Metric vector space):
A topological vector space having a distance function is called a **metric vector space**. ■

Example 1:
The topological vector space V with the distance function

$$\rho(\boldsymbol{x}, \boldsymbol{y}) = \begin{cases} 0 & \text{if } \boldsymbol{x} = \boldsymbol{y}, \\ 1 & \text{if } \boldsymbol{x} \neq \boldsymbol{y}, \end{cases} \tag{3.45}$$

[17] Also called a **metric**.

for arbitrary $\boldsymbol{x}, \boldsymbol{y} \in V$ is a metric vector space. ■

Example 2:
The set of real numbers $\mathbb{R}$ with the distance function $\rho(x,y) = |x-y|$ forms a metric vector space. ■

Example 3:
The set of ordered n-tuples of real numbers $\boldsymbol{x} = (x_1, x_2, \cdots, x_n)$ with the distance function

$$\rho(\boldsymbol{x}, \boldsymbol{y}) = \sqrt{\sum_{i=1}^{n} (x_i - y_i)^2} \tag{3.46}$$

is a metric vector space. This is in fact an n-dimensional Euclidean space, denoted by $\mathbb{R}^n$. ■

Example 4:
Consider again the set of ordered n-tuples of real numbers $\boldsymbol{x} = (x_1, x_2, \cdots, x_n)$ with an alternative distance function

$$\rho(\boldsymbol{x}, \boldsymbol{y}) = \max\left[|x_i - y_i|;\ 1 \le i \le n\right]. \tag{3.47}$$

The set also forms a metric vector space. ■

Comparison between Examples **3** and **4** tells us that the same topological vector space V can be metrized in different ways. These two examples call attention to the importance of distinguishing a pair (V, ρ) from V itself. Namely, a metric vector space (V, ρ) is a subset of a topological vector space V. But even so, our subsequent discussion focus mainly on metric vector spaces only; vector spaces without any distance function (or any topology) will be rarely dealt with. This is the reason why we henceforth call *metric* vector spaces just by "vector spaces" unless stated otherwise.

3.3.3 *Normed spaces*

A (metric) vector space V is said to be normed if to each element $\boldsymbol{x} \in V$ there is a corresponding non-negative number $\|\boldsymbol{x}\|$ which is called the **norm** of $\boldsymbol{x}$.

Definition (Normed space):
A (metric) vector space V equipped with a **norm** is called a **normed space**. The norm is defined as a real-valued function (denoted by $\|\ \|$) on a vector space V, which satisfies

1. $\|\lambda \boldsymbol{x}\| = |\lambda| \|\boldsymbol{x}\|$ for all $\lambda \in \mathbb{F}$ and $\boldsymbol{x} \in V$,
2. $\|\boldsymbol{x} + \boldsymbol{y}\| \le \|\boldsymbol{x}\| + \|\boldsymbol{y}\|$, and
3. $\|\boldsymbol{x}\| = 0$ if and only if $\boldsymbol{x} = \boldsymbol{0}$. ■

Obviously, a normed space is a (metric) vector space under the definition of the distance: $\rho(\boldsymbol{x},\boldsymbol{y}) = \|\boldsymbol{x}-\boldsymbol{y}\|$. Every norm determines a distance function, but not all distance function induce a norm.

Example 1:
The set of all n-tuples of real numbers: $\boldsymbol{x} = (x_1, x_2, \cdots, x_n)$ in which the norm is defined by

$$\|\boldsymbol{x}\| = \left(\sum_{i=1}^{n} x_i^p\right)^{1/p} \quad \text{with } p \geq 1. \tag{3.48}$$

This norm is referred to as the ℓ^p **norm.** ∎

Example 2:
We further obtain an alternative normed space if the norm of a vector $\boldsymbol{x} = (x_1, x_2, \cdots, x_n)$ is defined by

$$\|\boldsymbol{x}\| = \max\{|x_k|;\ 1 \leq k \leq n\}. \quad \blacksquare \tag{3.49}$$

Example 3:
The set of all continuous functions defined on the closed interval $[a,b]$, in which

$$\|f\| = \left[\int_a^b |f(x)|^p dx\right]^{1/p} \quad \text{with } p \geq 1 \tag{3.50}$$

is a normed space. This norm is called the L^p **norm.** ∎

Example 4:
An alternative normed space is obtained if the norm of the set of functions given by Example 3 is defined by

$$\|f\| \equiv \max\{|f(x)| : x \in [a,b]\}. \quad \blacksquare \tag{3.51}$$

3.3.4 Subspaces of normed spaces

Normed spaces contain the following two subclasses: the one is endowed with the completeness, and the other with the inner product. The former class, i.e., a class of complete normed vector spaces, are called **Banach spaces**. Here, the completeness of a space implies that every Cauchy sequence in the space is convergent. Refer to the arguments in section 3.2.5 for details.

Definition (Banach space):
If a normed space is complete, it is called a **Banach space.** ∎

Importantly, every *finite* dimensional normed space is necessarily complete, and thus being a Banach space. Conversely, *infinite* dimensional normed spaces may or may not be complete; therefore, not all of them are Banach spaces. This contrast emphasizes the essential difference between finite and infinite dimensional vector spaces.

Keypoint: In "finite" dimension, normed space are always complete. But in "infinite" dimension, normed spaces may or may not be complete.

Two of the most famous Banach spaces with *infinite* dimension are demonstrated below.

Example 1:
Suppose that a set of infinite-dimensional vectors $\boldsymbol{x} = (x_1, x_2, \cdots, x_n, \cdots)$ satisfies the condition:

$$\sum_{i=1}^{\infty} |x_i|^p < \infty, \quad (p \geq 1). \tag{3.52}$$

Then, this set is a Banach space, called an ℓ^p **space**, under the norm defined by

$$\|\boldsymbol{x}\|_p = \left(\sum_{i=1}^{\infty} |x_i|^p \right)^{1/p}. \tag{3.53}$$

The proof of their completeness will be given in section 4.1.2. ■

Example 2:
A set of functions $f(x)$ expressed by

$$\int_a^b |f(x)|^p dx < \infty \tag{3.54}$$

forms a Banach space, called an L^p **space**, under the norm:

$$\|f\|_p = \left(\int_a^b |f(x)|^p dx \right)^{1/p}. \tag{3.55}$$

The completeness will be proved in section 4.1.3. ■

Next we pay attention to the counterpart, *i.e.*, normed spaces that are not complete but endowed with inner product.

Definition (Pre-Hilbert space):
Inner product spaces that are not complete are called **pre-Hilbert spaces.** ■

Example:
Let $\{f_n\}$ be an infinite set of real-valued continuous functions on the closed interval $[0,1]$. The x-dependence of $f_n(x)$ is given by

$$f_n(x) = \begin{cases} 1, & \text{for } 0 \leq x \leq \frac{1}{2}, \\ 1 - 2n\left(x - \frac{1}{2}\right), & \text{for } \frac{1}{2} \leq x \leq \frac{1}{2}\left(1 + \frac{1}{n}\right), \\ 0, & \text{for } \frac{1}{2}\left(1 + \frac{1}{n}\right) \leq x \leq 1. \end{cases} \tag{3.56}$$

The graphs of $f_n(x)$ for $n = 1, 2, 10, 20$ are given in Fig. 3.8, which indicate that the infinite sequence $\{f_1, f_2, \cdots\}$ converges to a discrete step-wise function having a discontinuous jump at $x = 1/2$. For the sequence of the functions, we define the norm and inner product by

$$\|f_n\| = \sqrt{\int_0^1 [f_n(x)]^2 dx} \quad \text{and} \quad (f_m, f_n) = \int_0^1 f_m(x) f_n(x) dx. \tag{3.57}$$

Then the distance between two elements are written as

$$\begin{aligned} \|f_n - f_m\| &= \sqrt{\int_0^1 [f_n(x) - f_m(x)]^2 dx} \\ &= \left(1 - \frac{n}{m}\right) \sqrt{\frac{1}{6n}} \to 0 \quad \text{as } m, n \to \infty. \quad (m > n) \end{aligned}$$

Thus, the sequence $\{f_n\}$ is a Cauchy sequence. However, this sequence converges to the limit function

$$f(x) = \begin{cases} 1, & \text{if } 0 \le x \le \frac{1}{2}, \\ 0, & \text{if } \frac{1}{2} \le x \le 1, \end{cases} \tag{3.58}$$

which is not continuous and hence is not contained in the original set of functions. Consequently, the set of infinitely many functions $\{f_n\}$ given by (3.56) forms a non-complete inner product space, thus being a pre-Hilbert space. ■

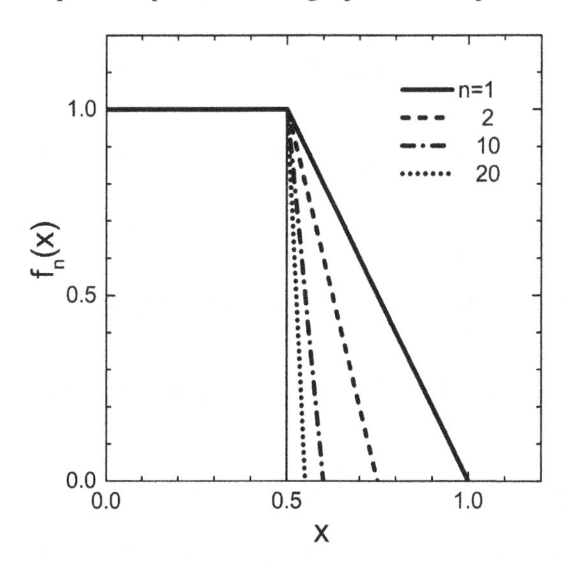

Figure 3.8: The function $f_n(x)$ given in (3.56). The infinite sequence $\{f_n(x)\}$ converges to a step function with a jump at $x = 1/2$.

3.3.5 Hilbert spaces

Eventually we are ready for enjoying the definition of Hilbert spaces. They are defined as the intersection between Banach spaces and pre-Hilbert spaces as stated below.

> **Definition (Hilbert space):**
> A complete normed space endowed with inner product is called a **Hilbert space**. ■

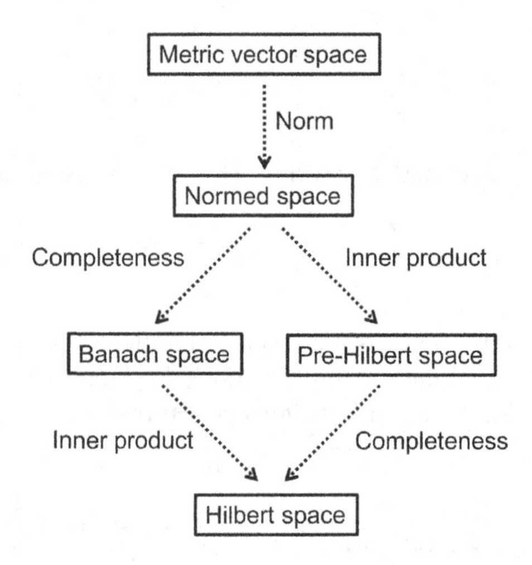

Figure 3.9: Roadmap from a metric vector space to a Hilbert space.

Example:
The ℓ^p spaces and L^p spaces with $p = 2$, so-called the ℓ^2 **spaces** and L^2 **spaces**, are Hilbert spaces. The inner product of each space is given by

$$(\boldsymbol{x}, \boldsymbol{y}) = \sum_{i=1}^{\infty} (x_i)^* y_i \text{ and } (f, g) = \int_a^b [f(x)]^* g(x) dx. \quad ■ \tag{3.59}$$

It should be emphasized that only the case of $p = 2$ among various ℓ^p spaces and L^p spaces form Hilbert spaces. This is because the spaces with $p \neq 2$ do not have inner product. Indeed, if the latter spaces were inner produce spaces, then the norms $\|\boldsymbol{x}\|_p$ and $\|f\|_p$ defined by (3.53) and (3.55), respectively, should have satisfied the parallelogram law given by (3.23). Namely, we must obtain the equalities:

$$\|\boldsymbol{x} + \boldsymbol{y}\|_p^2 + \|\boldsymbol{x} - \boldsymbol{y}\|_p^2 = 2\left(\|\boldsymbol{x}\|_p^2 + \|\boldsymbol{y}\|_p^2\right), \tag{3.60}$$

and

$$\|f+g\|_p^2+\|f-g\|_p^2=2\left(\|f\|_p^2+\|g\|_p^2\right). \tag{3.61}$$

However, the above-mentioned both fail unless $p = 2$. Accordingly, among the family of ℓ^p and L^p, only the spaces ℓ^2 and L^2 can afford to be a Hilbert space.

In summary, the difference between a Hilbert space and a Banach space is the source of the norm. In a Hilbert space, on the one hand, the norm is defined on the basis of an inner product. In a Banach space, on the other hand, the norm is defined directly without reference to an inner product.

Keypoint: ℓ^2 and L^2 spaces are Hilbert spaces, whereas ℓ^p and L^p spaces with $p \neq 2$ are Banach spaces.

Chapter 4

Hilbert Space

4.1 Basis and completeness

4.1.1 Basis of infinite-dimensional spaces

Every Hilbert space gives a means by which functions can be regarded as points in an infinite-dimensional space. The utility of this geometric perspective is found in our ability to generalize the notions of a length, orthogonality, and linear combinations of elements from elementary linear algebra to more abstract mathematical spaces. The key concept that governs this geometric perspective is a **basis** of the spaces. In the beginning of this chapter, we review the essential properties of a basis specifically in an infinite-dimensional space.

Given a vector space (either finite or infinite-dimensional), an orthonormal set of vectors works as a basis if it is **complete**. Once obtaining a basis, any vector in the space can be decomposed into a linear combination of orthonormal vectors. But there is one caution; when you try to construct a basis of an infinite-dimensional space from a set of infinitely many orthonormal vectors, you must check the completeness of the set of vectors. This caution comes from the fact that even if an infinite sum of vectors in a space is convergent, the limit may and may not be contained in the same vector space. Namely, the limit can deviate from the original space; in such a case, the set of vectors does no longer work as the basis of the infinite-dimensional space.

To look into this point, let us consider an infinite set $\{e_i\}$ $(i = 1,2,\cdots)$ of orthonormal vectors in a Hilbert space V. Next we choose one vector $\boldsymbol{x} \in V$, and then calculate the inner product of $\boldsymbol{e}_i$ and $\boldsymbol{x}$ expressed by

$$c_i = (\boldsymbol{e}_i, \boldsymbol{x}) \in \mathbb{C} \quad (i = 1,2,\cdots). \tag{4.1}$$

Using the first n terms of the infinite sequence $\{c_i\}$ and those of the infinite set of vectors $\{e_i\}$, we create a finite sum of the product $c_i e_i$ written by

$$\sum_{i=1}^{n} c_i e_i \equiv x_n. \tag{4.2}$$

For the pair of vectors x and x_n, **Schwarz's inequality** (3.21) gives

$$|(x, x_n)|^2 \le \|x\|^2 \cdot \|x_n\|^2 = \|x\|^2 \left(\sum_{i=1}^{n} |c_i|^2 \right). \tag{4.3}$$

On the other hand, taking the inner product of (4.2) with x yields

$$(x, x_n) = \sum_{i=1}^{n} c_i (x, e_i) = \sum_{i=1}^{n} |c_i|^2. \tag{4.4}$$

From (4.3) and (4.4), we have

$$\sum_{i=1}^{n} |c_i|^2 \le \|x\|^2. \tag{4.5}$$

Since (4.5) is true for arbitrarily large n, we obtain the following theorem.

Theorem (Bessel's inequality in infinite dimension):
Let $\{e_i\}$ $(i = 1, 2, \cdots)$ be an infinite set of orthonormal vectors in a Hilbert space V. Then for any $x \in V$ with $c_i = (e_i, x)$, we have

$$\sum_{i=1}^{\infty} |c_i|^2 \le \|x\|^2. \tag{4.6}$$

This inequality is referred to as **Bessel's inequality**. ■

The Bessel inequality shows that the limiting vector

$$\lim_{n\to\infty} x_n = \lim_{n\to\infty} \left(\sum_{i=1}^{n} c_i e_i \right) = \sum_{i=1}^{\infty} c_i e_i \tag{4.7}$$

has a finite norm represented by $\sum_{i=1}^{\infty} |c_i|^2$, thus being convergent. However, it remains unknown whether the limiting vector is identical to x. To make such a statement, the set $\{e_i\}$ should be complete as stated below.

Definition (Complete orthonormal vectors):
Let $\{e_i\}$ be an infinite set of orthonormal vectors in a Hilbert space V. The set is **complete** if there exist no vector in V (except for $\mathbf{0}$) that is orthogonal to all e_is. ■

> **Keypoint:** In an infinite-dimensional space, an orthonormal set $\{e_i\}$ can be a basis only when it is complete.

The following is an immediate consequence of the completeness of $\{e_i\}$.

Theorem (Parseval's identity):
Let $\{e_i\}$ be an infinite set of orthonormal vectors in a Hilbert space V. Then for any $\boldsymbol{x} \in V$,

$$\begin{aligned} & \{e_i\} \text{ is complete} \\ \Longleftrightarrow \quad & \|\boldsymbol{x}\|^2 = \sum_{i=1}^{\infty} |c_i|^2 \text{ with } c_i = (\boldsymbol{e}_i, \boldsymbol{x}) \quad \blacksquare \end{aligned} \tag{4.8}$$

Proof:
Let the set $\{e_i\}$ be complete, and consider the vector defined by

$$\boldsymbol{y} = \boldsymbol{x} - \sum_{i=1}^{\infty} c_i \boldsymbol{e}_i, \tag{4.9}$$

where $\boldsymbol{x} \in V$ and $c_i = (\boldsymbol{e}_i, \boldsymbol{x})$. It follows that for any $\boldsymbol{e}_j$,

$$(\boldsymbol{e}_j, \boldsymbol{y}) = (\boldsymbol{e}_j, \boldsymbol{x}) - \sum_{i=1}^{\infty} c_i (\boldsymbol{e}_j, \boldsymbol{e}_i) = c_j - \sum_{i=1}^{\infty} c_i \delta_{ji} = c_j - c_j = 0. \tag{4.10}$$

In view of the definition of the completeness of $\{e_i\}$, the result of $(\boldsymbol{e}_j, \boldsymbol{y}) = 0$ for any j means that $\boldsymbol{y}$ is the zero vector. Hence we have

$$\boldsymbol{x} = \sum_{i=1}^{\infty} c_i \boldsymbol{e}_i, \tag{4.11}$$

which implies

$$\|\boldsymbol{x}\|^2 = \sum_{i=1}^{\infty} |c_i|^2. \tag{4.12}$$

We now consider the converse. Suppose that $\boldsymbol{x}$ be orthogonal to all e_i, which means

$$(\boldsymbol{e}_i, \boldsymbol{x}) = c_i = 0 \text{ for all } i. \tag{4.13}$$

It follows from (4.8) and (4.13) that $\|\boldsymbol{x}\|^2 = 0$, which in turn gives $\boldsymbol{x} = \mathbf{0}$. This completes the proof. $\blacksquare$

The above discussion tells us that the completeness is a necessary and sufficient condition for an infinite set of orthonormal vectors $\{e_i\}$ to be a basis of an infinite-dimensional space. Emphasis should be placed on the fact that only the presence of infinitely many vectors is not sufficient; the completeness of the set of vectors should be indispensable.

Keypoint: Parseval's identity provides a criterion for the availability of an infinite set of vectors $\{e_i\}$ as a basis.

4.1.2 Completeness of ℓ^2 spaces

In the previous chapter, we observed that ℓ^2 spaces and L^2 spaces are the two most important classes of Hilbert spaces. Among the two, the completeness property of the ℓ^2 space on the field $\mathbb{F}$ can be examined by the following procedure.

In general, the completeness of a given vector space V is characterized by the fact that every Cauchy sequences $(\boldsymbol{x}_n)$ involved in the space converges to an element $\boldsymbol{x} \in V$ such that

$$\lim_{n\to\infty} \|\boldsymbol{x} - \boldsymbol{x}_n\| = 0 \tag{4.14}$$

To prove the completeness of the ℓ^2 space, therefore, we will show in turn that:

i) Every Cauchy sequence $(\boldsymbol{x}_n)$ in the ℓ^2 space converges to a limit $\boldsymbol{x}$, and
ii) The limit $\boldsymbol{x}$ belongs to the ℓ^2 space.

First we work on the statement **i)** Consider a set of infinite-dimensional vectors

$$\boldsymbol{x}^{(n)} = \left(x_1^{(n)}, x_2^{(n)}, \cdots\right) \tag{4.15}$$

wherein $x_i^{(n)} \in \mathbb{F}$ with $\mathbb{F} = \mathbb{R}$ or $\mathbb{C}$, and let the sequence of vectors $\left\{\boldsymbol{x}^{(1)}, \boldsymbol{x}^{(2)}, \cdots\right\}$ be a Cauchy sequence in the sense of the norm

$$\|\boldsymbol{x}\| = \sqrt{\sum_{i=1}^{\infty} |x_i|^2} < \infty. \tag{4.16}$$

Then, for any $\varepsilon > 0$, there exists an integer N such that

$$m, n > N \Rightarrow \left\|\boldsymbol{x}^{(m)} - \boldsymbol{x}^{(n)}\right\| = \sqrt{\sum_{i=1}^{\infty} \left|x_i^{(m)} - x_i^{(n)}\right|^2} < \varepsilon. \tag{4.17}$$

Using the relation of

$$\sqrt{\sum_{i=1}^{\infty} \left|x_i^{(m)} - x_i^{(n)}\right|^2} \le \sum_{i=1}^{\infty} \left|x_i^{(m)} - x_i^{(n)}\right|, \tag{4.18}$$

we see from (4.17) that

$$\left|x_i^{(m)} - x_i^{(n)}\right| < \varepsilon \tag{4.19}$$

for every i and every $m, n > N$. Furthermore, since (4.17) is true in the limit $m \to \infty$, we find

$$\left\|\boldsymbol{x} - \boldsymbol{x}^{(n)}\right\| < \varepsilon \tag{4.20}$$

for arbitrary $n > N$. The inequalities (4.19) and (4.20) mean that $\boldsymbol{x}^{(n)}$ converges to the limiting vector expressed by $\boldsymbol{x} \equiv (x_1, x_2, \cdots)$, in which the component $x_i \in \mathbb{F}$ is defined by

$$x_i = \lim_{n\to\infty} x_i^{(n)}. \tag{4.21}$$

(That the limit (4.21) belongs to $\mathbb{F}$ is guaranteed by the completeness of $\mathbb{F}$.)

The remained task is to show the statement **ii)**, namely that the limiting vector $\boldsymbol{x}$ belongs to the original space ℓ^2. By the **triangle inequality**, we have

$$\|\boldsymbol{x}\| = \left\|\boldsymbol{x} - \boldsymbol{x}^{(n)} + \boldsymbol{x}^{(n)}\right\| \leq \left\|\boldsymbol{x} - \boldsymbol{x}^{(n)}\right\| + \left\|\boldsymbol{x}^{(n)}\right\|. \tag{4.22}$$

Hence for every $n > N$ and for every $\varepsilon > 0$, we obtain

$$\|\boldsymbol{x}\| < \varepsilon + \left\|\boldsymbol{x}^{(n)}\right\|. \tag{4.23}$$

Because the Cauchy sequence $(\boldsymbol{x}^{(1)}, \boldsymbol{x}^{(2)}, \cdots)$ is bounded, $\|\boldsymbol{x}\|$ cannot be greater than

$$\varepsilon + \limsup_{n\to\infty} \left\|\boldsymbol{x}^{(n)}\right\|, \tag{4.24}$$

and is therefore finite. This implies that the limit vector $\boldsymbol{x}$ belongs to ℓ^2 on $\mathbb{F}$. Consequently, we have proved that the ℓ^2 space on $\mathbb{F}$ is complete.

Among various kinds of Hilbert spaces, the ℓ^2 space has a significant importance in mathematical physics, mainly because it serves a ground for the theory of quantum mechanics. In fact, any element $\boldsymbol{x}$ of the ℓ^2 space works as a possible state vector of quantum objects if it satisfies the normalized condition

$$\|\boldsymbol{x}\| = \sum_{i=1}^{\infty} |x_i|^2 = 1. \tag{4.25}$$

4.1.3 Completeness of L^2 spaces

Here is given a short comment on the completeness of L^2 **spaces**. An L^2 space is spanned by **square integrable**[1] functions $\{f_n\}$ with respect to x on a closed interval, say $[a,b]$. To prove the completeness of the L^2 space, it is necessary to show that every Cauchy sequence $\{f_n\}$ in the L^2 space converges to a limit function f, and then verify that the f belongs to L^2.

In the proof, we shall manipulate integrations of infinite sums of functions and the exchange between limit procedures and integration procedures. But these mathematical manipulations sometimes fail in a realm of elementary calculus; they require the techniques derived from the **Lebesgue integral theory**, which will be explained in Chapter 6. This is the reason why we omit at this stage to describe the proof of the completeness of L^2 spaces and postpone it to section 6.5.4.

4.1.4 Mean convergence in L^2 spaces

It is interesting that the completeness of L^2 spaces induces a new class of convergence for infinite sets of functions. Observe that the completeness of L^2 spaces is rephrased

[1] A function $f(x)$ is said to be **square integrable** on X if $\int_X |f(x)|^2 dx$ is finite.

by the following sentence: For any small $\varepsilon > 0$, it is possible to find N such that

$$n > N \Rightarrow \|f - f_n\| < \varepsilon \text{ with } f_n, f \in L^2. \tag{4.26}$$

Hence, we can say that the infinite sequence $\{f_n\}$ converges to f in the norm of the L^2 space. Notably, the convergence of the type (4.26), represented by $\|f_n - f\| = 0$ in the sense of L^2 norm, relies on the notion of **the convergence in the mean** or known as **the mean convergence**.

The mean convergence is inherently different from an elementary class of convergences such as the pointwise convergence[2]. and the uniform convergence.[3] A primary feature of the mean convergence is that the quantitative deviation between $f_n(x)$ and $f(x)$ is measured not by the difference $f(x) - f_n(x)$, but by the norm in the L^2 space based on the integration procedure:

$$\|f - f_n\| = \sqrt{\int_a^b |f(x) - f_n(x)|^2 dx}. \tag{4.27}$$

Hence, when $f(x)$ is convergent in the mean to $f_n(x)$ on the interval $[a, b]$, there may exist a finite number of **isolated points** such that $f(x) \neq f_n(x)$. Obviously, this situation is not allowed in neither case of uniform nor pointwise convergence.

When $f(x)$ converges to $g(x)$ in the mean, it is conventional to say:

"$f(x)$ is equal to $g(x)$ **almost everywhere**".

A rigorous definition of the equality almost everywhere is based on the Lebesgue integral theory, as we will see in section 6.3.4.

Keypoint: Mean convergence of $\{f_n\}$ to f guarantees the equality between $\lim_{n\to\infty} f_n$ and f almost everywhere, not totally everywhere.

4.2 Equivalence of L^2 spaces with ℓ^2 spaces

4.2.1 Generalized Fourier coefficient

Having clarified the completeness property of the two specific Hilbert spaces ℓ^2 and L^2, we introduce the two important concepts: **generalized Fourier coefficients**

[2]Pointwise convergence of a sequence $\{f_n\}$ to f means that $\lim_{n\to\infty} f_n(x) = f(x)$ for every x in a domain considered.

[3]**Uniform convergence** of a sequence $\{f_n\}$ to f means that $\lim\sup_{n\to\infty}\{|f_n(x) - f(x)|\} = 0$ for every x in a domain considered. This is a stronger statement than the assertion of **pointwise convergence**. In fact, every uniformly convergent sequence is pointwise convergent to the same limit, but some pointwise convergent sequences are not uniformly convergent. For example, we have $\lim_{n\to\infty} x^n$ converges pointwise to 0 on the interval $[0, 1)$, but not uniformly on the interval $[0, 1)$.

and **generalized Fourier series**. Readers may be familiar with the Fourier series associated with trigonometric functions written by

$$f(x) = \frac{a_0}{2} + \sum_{k=1}^{\infty} (a_k \cos kx + b_k \sin kx), \tag{4.28}$$

or the one with imaginary exponentials written by

$$f(x) = \sum_{k=-\infty}^{\infty} c_k e^{ikx}. \tag{4.29}$$

Notably, the concepts of Fourier series and Fourier coefficients can be extended to more general concepts as stated below.

Definition (Generalized Fourier series):
Consider a set of square integrable functions $\{\phi_i\}$ that are orthonormal (not necessarily complete) in the norm of the L^2 space. Then the numbers

$$c_k = (f, \phi_k) \tag{4.30}$$

are called the **Fourier coefficients** of the function $f \in L^2$ relative to the orthonormal set $\{\phi_i\}$, and the series

$$\sum_{k=1}^{\infty} c_k \phi_k \tag{4.31}$$

is called the **Fourier series** of f with respect to the set $\{\phi_i\}$. ■

In general, Fourier series (4.31) may and may not be convergent; its convergence property is determined by the features of the functions f and the associated orthonormal set of functions $\{\phi_k\}$.

The importance of the Fourier coefficients (4.30) becomes apparent by seeing that they consist of the ℓ^2 space. In fact, since c_k is the inner product of f and ϕ_k, it yields the **Bessel's inequality** in terms of c_k and f such as

$$\sum_{k=1}^{\infty} |c_k|^2 \leq \|f\|. \tag{4.32}$$

From the hypothesis of $f \in L^2$, the norm $\|f\|$ remains finite. Hence, the inequality (4.32) assures the convergence of the infinite series $\sum_{k=1}^{\infty} |c_k|^2$. This convergence means that the sequence of Fourier coefficients $\{c_k\}$ is an element of the space ℓ^2, whichever orthonormal set of functions $\phi_k(x)$ we choose. In this context, the two elements:

$$f \in L^2 \quad \text{and} \quad \boldsymbol{c} = (c_1, c_2, \cdots) \in \ell^2 \tag{4.33}$$

are connected via the Fourier coefficient (4.30).

Keypoint: Generalized Fourier coefficients make clear a close relationship between the two distinct Hilbert spaces ℓ^2 and L^2.

4.2.2 Riesz-Fisher theorem

Recall that every Fourier coefficients satisfy Bessel's inequality (4.32). Hence, in order for a given set of complex numbers (c_i) to constitute the Fourier coefficients of a function $f \in L^2$, it is necessary that the series

$$\sum_{k=1}^{\infty} |c_k|^2 \tag{4.34}$$

converges. As a matter of fact, this condition is not only necessary, but also sufficient as stated in the theorem below.

Theorem (Riesz-Fisher theorem):
Given any set of complex numbers (c_i) such that

$$\sum_{k=1}^{\infty} |c_k|^2 < \infty, \tag{4.35}$$

there exists a function $f \in L^2$ such that

$$c_k = (f, \phi_k) \quad \text{and} \quad \sum_{k=1}^{\infty} |c_k|^2 = \|f\|^2, \tag{4.36}$$

where $\{\phi_i\}$ is a complete orthonormal set. ■

Proof:
Set a linear combination of $\phi_k(x)$ as

$$f_n(x) = \sum_{k=1}^{n} c_k \phi_k(x), \tag{4.37}$$

where c_k are arbitrary complex numbers satisfying the condition (4.35). Then, for a given integer $p \geq 1$, we obtain

$$\|f_{n+p} - f_n\|^2 = \|c_{n+1}\phi_{n+1} + \cdots + c_{n+p}\phi_{n+p}\|^2 = \sum_{k=n+1}^{n+p} |c_k|^2. \tag{4.38}$$

Let $p = 1$ and $n \to \infty$. Then, due to the condition (4.35), we have

$$\|f_{n+1} - f_n\| = |c_{n+1}|^2 \to 0. \ (n \to \infty) \tag{4.39}$$

This tells us that the infinite sequence $\{f_n\}$ defined by (4.37) always converges in the mean to a function $f \in L^2$.

Our remaining task is to show that this limit function f satisfies the condition (4.36). Let us consider the inner product:

$$(f, \phi_i) = (f_n, \phi_i) + (f - f_n, \phi_i), \tag{4.40}$$

where $n \geq i$ is assumed. It follows from (4.37) that the first term on the right hand side is equal to c_i. The second term vanishes as $n \to \infty$, since

$$|(f - f_n, \phi_i)| \leq \|f - f_n\| \cdot \|\phi_i\| \to 0, \quad (n \to \infty) \tag{4.41}$$

where the mean convergence property of $\{f_n\}$ to f was used. In addition, the left hand side of (4.40) is independent of n. Hence, taking the limit $n \to \infty$ in both sides of (4.40), we obtain

$$(f, \phi_i) = c_i. \tag{4.42}$$

This indicates that c_i is the Fourier coefficient of f relative to ϕ_i. From our assumption, the set $\{\phi_i\}$ is complete and orthonormal. Hence, the Fourier coefficients (4.42) satisfy **Parseval's identity**:

$$\sum_{k=1}^{\infty} |c_k|^2 = \|f\|^2. \tag{4.43}$$

The results (4.42) and (4.43) are identical to the condition (4.36), thus the theorem has been proved. ■

4.2.3 Isomorphism between ℓ^2 and L^2

An immediate consequence of **Riesz-Fisher theorem** previously mentioned is the **isomorphism** between the Hilbert spaces L^2 and ℓ^2. An isomorphism is a one-to-one correspondence that preserves all algebraic structure. For instance, two vector spaces U and V (over the same number field) are isomorphic if there exists a one-to-one correspondence between the vectors $\boldsymbol{x}_i$ in U and $\boldsymbol{y}_i$ in V, say $\boldsymbol{y}_i = F(\boldsymbol{x}_i)$ such that

$$F(\alpha_1 \boldsymbol{x}_1 + \alpha_2 \boldsymbol{x}_2) = \alpha_1 F(\boldsymbol{x}_1) + \alpha_2 F(\boldsymbol{x}_2). \tag{4.44}$$

Let us prove the above issue. Choose an arbitrary complete orthonormal set $\{\phi_n\}$ in L^2 and assign to each function $f \in L^2$ the sequence $(c_1, c_2, \cdots, c_n, \cdots)$ of its Fourier coefficients with respect to this set. Since

$$\sum_{k=1}^{\infty} |c_k|^2 = \|f\|^2 < \infty, \tag{4.45}$$

the sequence $(c_1, c_2, \cdots, c_n, \cdots)$ is an element of ℓ^2. Conversely, in view of Riesz-Fisher theorem, for every element $(c_1, c_2, \cdots, c_n, \cdots)$ of ℓ^2 there is a function $f(x) \in L^2$ whose Fourier coefficients are $c_1, c_2, \cdots, c_n, \cdots$. This correspondence between the elements of L^2 and ℓ^2 is one-to-one. Furthermore, if

$$f \longleftrightarrow \{c_1, c_2, \cdots, c_n, \cdots\} \tag{4.46}$$

$$\text{and} \quad g \longleftrightarrow \{d_1, d_2, \cdots, d_n, \cdots\}, \tag{4.47}$$

then

$$f + g \longleftrightarrow \{c_1 + d_1, \cdots, c_n + d_n, \cdots\} \tag{4.48}$$

$$\text{and} \quad kf \longleftrightarrow \{kc_1, kc_2, \cdots, kc_n, \cdots\}, \tag{4.49}$$

which readily follow from the definition of Fourier coefficients. That is, addition and multiplication by scalars are preserved by the correspondence. Furthermore, in view of Parseval's identity, it follows that[4]

$$(f,g) = \sum_{i=1}^{\infty} c_i^* d_i. \tag{4.50}$$

All of these facts assure the isomorphism between the spaces L^2 and ℓ^2, i.e., the one-to-one correspondence between the elements of L^2 and ℓ^2 that preserves the algebraic structures of the space. In this context, it is possible to say that every element $\{c_i\}$ in ℓ^2 spaces serve as a *coordinate system* of the L^2 space, and vice versa.

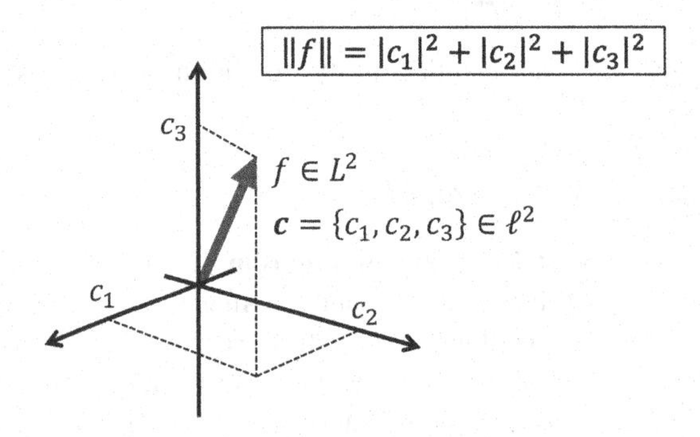

Figure 4.1: Analogy of geometric arrow representation for the isomorphism between L^2 and ℓ^2.

The isomorphism between L^2 and ℓ^2 is closely related to the theory of quantum mechanics. Quantum mechanics originally consisted of two distinct theories: **Heisenberg's matrix mechanics** based on infinite-dimensional vectors, and **Schrödinger's wave mechanics** based on square integrable functions. From the mathematical point of view, the difference between two theories is reduced to the fact that the former theory used the space ℓ^2, while the latter used the space L^2. Hence, the isomorphism between the two spaces verify the equivalence of two theories describing the nature of quantum mechanics.

[4]This equality is verified because of the relations: $(f,f) = \sum_{i=1}^{\infty} |c_i|^2$ and $(g,g) = \sum_{i=1}^{\infty} |d_i|^2$, and their consequences:

$$\begin{aligned}(f+g, f+g) &= (f,f) + 2(f,g) + (g,g) = \sum_{i=1}^{\infty} |c_i + d_i|^2 \\ &= \sum_{i=1}^{\infty} |c_i|^2 + 2\sum_{i=1}^{\infty} c_i^* d_i + \sum_{i=1}^{\infty} |d_i|^2.\end{aligned}$$

Chapter 5

Tensor Space

5.1 Two faces of one tensor

5.1.1 Tensor in practical science

In view of physics and engineering, a **tensor** is regarded to be a collection of index quantities; we thus often say that tensors are generalizations of **scalars** (that have no indices), **vectors** (exactly one index), and **matrices** (exactly two indices), to an arbitrary number of indices. Indeed, the notation for a tensor (i.e., $\boldsymbol{A} = [A_{ijk\ell m\cdots}]$) is similar to that of a matrix (i.e., $\boldsymbol{A} = [A_{ij}]$), except that a tensor may have an arbitrary number of indices.

Tensors provide a mathematical framework for formulating and solving problems in areas of electromagnetics, elasticity, fluid mechanics, and general relativity. A typical example of the use of tensors in electromagnetics has to do with the magnetic flux density $\boldsymbol{B}$. For free space, the magnetic flux density is related to the magnetization $\boldsymbol{H}$ through the permeability μ by the expression

$$\boldsymbol{B} = \mu \boldsymbol{H}, \quad \text{or equivalently} \quad B_i = \mu H_i \;\; (i = x, y, z), \tag{5.1}$$

where μ is a scalar with a specific value. Since μ is a scalar, $\boldsymbol{B}$ and $\boldsymbol{H}$ in free space possess the same direction. In some exotic materials, however, these two vectors differ in direction due to peculiar electrical characteristics of component atoms or molecules. In such materials, the scalar permeability μ should be replaced by the tensor permeability $\underline{\mu} = [\mu_{ij}]$ having two indices, and the above equation is rewritten as

$$\boldsymbol{B} = \underline{\mu} \cdot \boldsymbol{H} \quad \text{or} \quad B_i = \mu_{ij} H_j \;\; (i, j = x, y, z). \tag{5.2}$$

In this case, the tensor $\underline{\mu}$ plays a role of relating the direction of $\boldsymbol{B}$ to that of $\boldsymbol{H}$, which is the reason why components of the tensor μ_{ij} are labeled by two indices.

Another important example of tensors used in practice is a stress tensor that appears in the theory of continuum mechanics.[1] As a terminology in mechanics, stress is defined by the internal force per unit area that an infinitesimal portion of an elastic medium exerts on an adjacent portion of the medium. We know that force is a vector. Also, area can be represented as a vector by associating it with a direction, i.e., the differential area $d\boldsymbol{S}$ is a vector with magnitude $dS = |d\boldsymbol{S}|$ and direction normal to the area element. The vector representation of the area is required because in general elastic objects, an applied force induces two different types of stress: tensile stress (normal force) and shear stress (tangential force). Hence, the resulting force $d\boldsymbol{F}$ due to the stress $\underline{\boldsymbol{\sigma}} = [\sigma_{ij}]$ acting on a differential surface element $d\boldsymbol{S}$ is given by

$$d\boldsymbol{F} = \underline{\boldsymbol{\sigma}} \cdot d\boldsymbol{S}, \quad \text{or} \quad F_i = \sigma_{ij} S_j \;\; (i, j = x, y, z). \tag{5.3}$$

In this case, too, the stress tensor $\underline{\boldsymbol{\sigma}}$ relates the directions of the two vectors, $\boldsymbol{F}$ and $d\boldsymbol{S}$.

> **Keypoint:** In elementary sense, a tensor is a "*collection*" of indexed quantities.

5.1.2 *Tensor in abstract mathematics*

We have just reviewed that in an elementary sense, tensors are defined as indexed quantities. A tensor having two indices transforms a vector to the other vector in a coordinate system. In the example above, the tensor $\underline{\boldsymbol{\sigma}} = [\sigma_{ij}]$ transforms the vector $d\boldsymbol{S}$ to $d\boldsymbol{F}$ in the three-dimensional Euclidean space spanned by x, y, z axes. Notice that this definition of tensors relies on the coordinate representation of vectors, through which the length and orientation of the vectors are described. In the following discussion, we introduce an alternative definition of tensors; it is free from specifying the coordinate system, thus being suitable for describing geometric properties of abstract vector spaces.

To begin with, we assume that a tensor is not as a set of index quantities but a **linear function** (or **linear mapping**) acting on vector spaces. For instance, a second-order tensor $\boldsymbol{T}$ is identified with a linear function that associates two vectors $\boldsymbol{v}$ and $\boldsymbol{w}$ to a real number $c \in \mathbb{R}$, which is symbolized by

$$\boldsymbol{T}(\boldsymbol{v}, \boldsymbol{w}) = c. \tag{5.4}$$

Emphasis should be placed on the fact that such generalized definition of tensors applies to all kinds of general vector spaces, no matter which spaces are endowed with the notions of **distance**, **norm**, or **inner product**. In contrast, tensors in a realm of elementary linear algebra applies only to the three-dimensional Euclidean space,

[1]The word "tensor" is originally derived from the Latin "tensus" meaning stress or tension in an elastic medium.

a particular class of inner product spaces. We shall see that, however, the concept of tensor can be extended beyond the inner product spaces, by introducing a more general definition outlined above.

> **Keypoint:** In functional analysis, a tensor is a "*mapping*" that converts a set of vectors into a number.

Throughout this chapter, we restrict our arguments to finite-dimensional vector spaces over $\mathbb{R}$ in order to give a minimum course for general tensor calculus.

5.2 "Vector" as a linear function

5.2.1 Linear function spaces

Some preliminary notions are required to give a generalized tensor; these notions include **linear function space**, **dual space**, **direct product**, **tensor product**, etc. We will learn these concepts step-by-step in order to reach the rigorous definition of tensors given by section 5.3.4.

Let us begin with a short review on the nature of **linear functions**. Given two vector spaces V and W, it is possible to set a function f such as

$$f : V \to W. \tag{5.5}$$

In particular, the function f is called a **linear function** (or **linear mapping**) of V into W if it yields for all $\boldsymbol{v}_1, \boldsymbol{v}_2 \in V$ and $c \in \mathbb{R}$,

$$\begin{aligned} f(\boldsymbol{v}_1 + \boldsymbol{v}_2) &= f(\boldsymbol{v}_1) + f(\boldsymbol{v}_2), \\ f(c\boldsymbol{v}_1) &= cf(\boldsymbol{v}_1). \end{aligned}$$

In elementary calculus, the concepts of vectors and linear functions are distinguished from each other; vectors are elements of a vector space, and linear functions provide a correspondence between them. However, in view of the vector axioms listed in section 3.1.2, we observe that the set of linear functions $f, g, \cdots$ of V into W also forms a vector space. In fact, **addition** and **scalar multiplication** for the linear functions can be defined by

$$(f+g)\boldsymbol{v} = f(\boldsymbol{v}) + g(\boldsymbol{v}), \tag{5.6}$$

and

$$(cf)\boldsymbol{v} = cf(\boldsymbol{v}), \tag{5.7}$$

where $\boldsymbol{v} \in V$ and $f(\boldsymbol{v}), g(\boldsymbol{v}) \in W$. This fact poses an idea that we construct a vector space, denoted by $\mathscr{L}(V, W)$, which is spanned by a set of linear functions f such as

$$f : V \to W. \tag{5.8}$$

It is trivial to verify that $f+g$ and cf are again linear functions, so that they belong to the same vector space $\mathscr{L}(V,W)$. These arguments are summarized by the following important theorem:

Theorem:
Let V and W be vector spaces. A set of linear functions $f: V \to W$ forms a vector space, denoted by $\mathscr{L}(V,W)$. ∎

> **Definition (Linear function space):**
> A vector space $\mathscr{L}(V,W)$ is called a **linear function space**. ∎

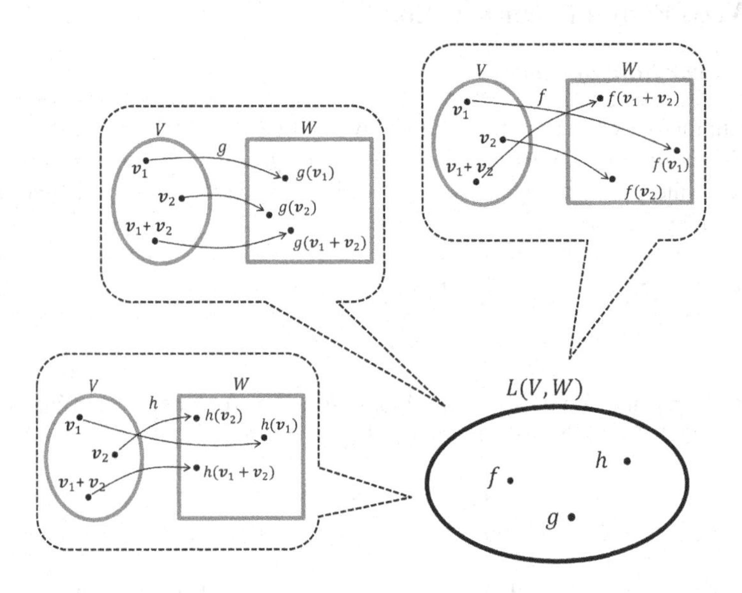

Figure 5.1: Linear function space $\mathscr{L}(V,W)$ is a vector space that consists of linear functions $f, g, h, \cdots$. All the linear functions of $f, g, h, \cdots$ map an element of V onto an element of W.

We have just learned that linear functions $f_1, f_2, \cdots$ of V into W are elements of a vector space $\mathscr{L}(V,W)$. This fact is analogous to that vectors $\boldsymbol{v}_1, \boldsymbol{v}_2, \cdots$ are elements of a vector space V. Importantly, the analogy implies that a linear function $f \in \mathscr{L}(V,W)$ can be regarded as a *vector*, and conversely, that a vector $\boldsymbol{v} \in V$ can be regarded as a *linear function*. Such identification of vectors and linear functions is crucially important for our quest to master the functional analysis. We will revisit this identification when introducing the notion of **natural pairing**, which will be demonstrated in section 5.2.3.

5.2.2 *Dual spaces*

Let V^* denote a set of all linear functions such as

$$\tau : V \to \mathbb{R}. \tag{5.9}$$

(Note that the asterisk attached to V^* does *not* mean complex conjugate.) Then, since

$$V^* = \mathscr{L}(V, \mathbb{R}), \tag{5.10}$$

it follows that V^* is a vector space. The vector space V^* is called the **dual space** (or **conjugate space**) of V, whose elements $\tau \in V^*$ associates a vector $\boldsymbol{v} \in V$ to a real number $c \in \mathbb{R}$, as symbolized by

$$\tau(\boldsymbol{v}) = c. \tag{5.11}$$

Definition (Dual space):
The vector space $V^* \equiv \mathscr{L}(V, \mathbb{R})$ is called a **dual space** of V. ■

Particularly important elements of V^* are linear functions

$$\varepsilon^i : V \to \mathbb{R} \quad (i = 1, 2, \cdots, n) \tag{5.12}$$

that associate a **basis** vector $\boldsymbol{e}_i \in V$ to the unit number 1. In fact, a set of such linear functions $\{\varepsilon^j\}$ serves as a basis of the dual space V^* as stated below.

Definition (Dual basis):
For each basis $\{\boldsymbol{e}_i\}$ for V, there is a unique basis $\{\varepsilon^j\}$ for V^* such that[2]

$$\varepsilon^j(\boldsymbol{e}_i) = \delta_i^j. \tag{5.13}$$

The linear functions $\varepsilon^j : V \to \mathbb{R}$ defined by (5.13) are called the **dual basis** to the basis $\{\boldsymbol{e}_i\}$ of V. ■

In the statement above, it is intentional to raise the index j attached to ε^j. This convention is necessary for providing a consistent notation of components of generalized tensors that shall be demonstrated in section 5.4.

Keypoint: The dual basis $\{\varepsilon^j\}$ is a looking-glass of the basis $\boldsymbol{e}_i$.

[2]Here and henceforth, we use the **Einstein's summation convention**. According to the convention, when an index variable appears twice in a single term, it implies summation of that term over all the values of the index. For instance, the summation $y = \sum_{i=1}^{\infty} c_i x^i = c_1 x^1 + c_2 x^2 + c_3 x^3 + \cdots$ is reduced by the convention to $y = c_i x^i$. The upper indices are not exponents but are indices of coordinates.

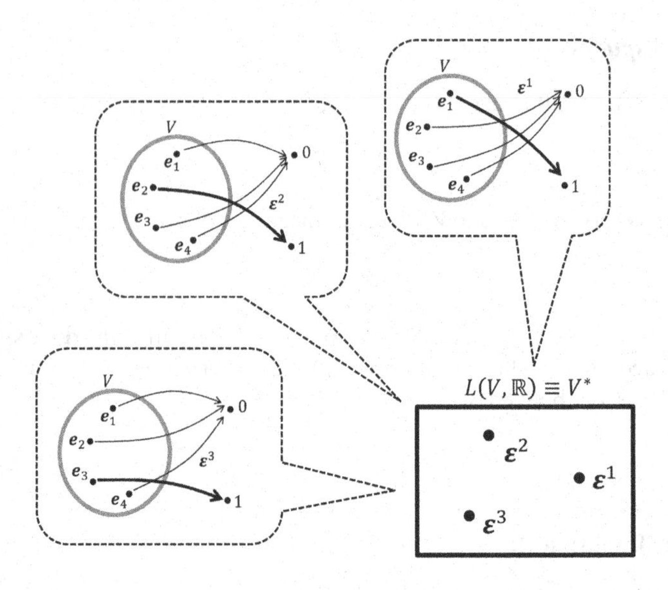

Figure 5.2: Dual space V^* is a vector space that consists of linear functions $\varepsilon^1, \varepsilon^2, \cdots$.

Example:
Expand a vector $\boldsymbol{v} \in V$ as

$$\boldsymbol{v} = v^i \boldsymbol{e}_i, \tag{5.14}$$

to find that

$$\varepsilon^j(\boldsymbol{v}) = \varepsilon^j\left(v^i \boldsymbol{e}_i\right) = v^i \varepsilon^j\left(\boldsymbol{e}_i\right) = v^i \delta_i^j = v^j. \tag{5.15}$$

This indicates that ε^j is the linear function that reads off the jth component of $\boldsymbol{v}$ with respect to the basis $\{\boldsymbol{e}_i\}$. It is symbolically written by

$$\varepsilon^j : \boldsymbol{v} \to v_j \quad (j = 1, 2, \cdots, n) \quad \blacksquare \tag{5.16}$$

5.2.3 Equivalence between vectors and linear functions

If V is a vector space and $\tau \in V^*$, then τ is a function of the variable $\boldsymbol{v} \in V$ that generates a real number denoted by $\tau(\boldsymbol{v})$. Due to the identification of vectors and linear functions, however, it is possible to twist our viewpoint around and consider $\boldsymbol{v}$ as a *function* of the *variable* τ, with the real value $\boldsymbol{v}(\tau) = \tau(\boldsymbol{v})$ again. When we take this latter viewpoint, $\boldsymbol{v}$ is a linear function on V^*.

The two viewpoints contrasted above appears to be asymmetric. This asymmetry can be eliminated by introducing the notation:

$$\langle \ , \ \rangle : V \times V^* \to \mathbb{R}, \tag{5.17}$$

which gives

$$\langle \boldsymbol{v}, \tau \rangle = \tau(\boldsymbol{v}) = \boldsymbol{v}(\tau) \in \mathbb{R}. \tag{5.18}$$

Here, $\langle \ , \ \rangle$ is a function of two variables $\boldsymbol{v}$ and $\boldsymbol{\tau}$, called the **natural pairing** of V and V^* into $\mathbb{R}$. It is easily verified that $\langle \ , \ \rangle$ is bilinear (see section 5.3.2).

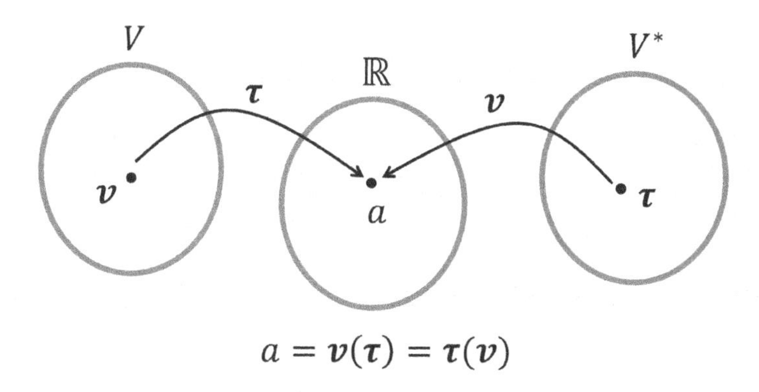

Figure 5.3: Natural pairing $\langle \ , \ \rangle$ is a linear function that transforms simultaneously two elements $\boldsymbol{v} \in V$ and $\boldsymbol{\tau} \in V^*$ to a real number $\langle \boldsymbol{v}, \boldsymbol{\tau} \rangle = \boldsymbol{\tau}(\boldsymbol{v}) = \boldsymbol{v}(\boldsymbol{\tau})$.

Keypoint: Every vector is a linear function. Every linear function is a vector.

5.3 Tensor as a multilinear function

5.3.1 Direct product of vector spaces

To achieve the desired new definition of tensors, three more concepts are required; they include **direct product** of vector spaces, **multilinearity**, and **tensor product**. We will address each of them in turn through this section.

The first to be considered is the direct product of vector spaces. Let V and W be vector spaces, then we can establish a new vector space by forming the **direct product** (or **Cartesian product**) $V \times W$ of the two spaces. The direct product $V \times W$ consists of *ordered* pairs $(\boldsymbol{v}, \boldsymbol{w})$ with $\boldsymbol{v} \in V$ and $\boldsymbol{w} \in W$, as symbolized by

$$V \times W = \{(\boldsymbol{v}, \boldsymbol{w}) \mid \boldsymbol{v} \in V, \boldsymbol{w} \in W\}. \tag{5.19}$$

The addition and scalar multiplication for the elements are defined by[3]

$$\begin{aligned}(\boldsymbol{v},\boldsymbol{w}_1)+(\boldsymbol{v},\boldsymbol{w}_2) &= (\boldsymbol{v},\boldsymbol{w}_1+\boldsymbol{w}_2), &(5.20)\\ (\boldsymbol{v}_1,\boldsymbol{w})+(\boldsymbol{v}_2,\boldsymbol{w}) &= (\boldsymbol{v}_1+\boldsymbol{v}_2,\boldsymbol{w}), &(5.21)\\ c(\boldsymbol{v},\boldsymbol{w}) &= (c\boldsymbol{v},c\boldsymbol{w}). &(5.22)\end{aligned}$$

The dimension of the resulting vector spaces $V \times W$ equals the product of the dimensions of V and W.

5.3.2 *Multilinear functions*

Let V_1, V_2 and W be vector spaces. A function

$$f: V_1 \times V_2 \to W \tag{5.23}$$

is called **bilinear** if it is linear in each variable; that is,

$$\begin{aligned}f(a\boldsymbol{v}_1+b\boldsymbol{v}'_1,\, \boldsymbol{v}_2) &= af(\boldsymbol{v}_1,\boldsymbol{v}_2)+bf(\boldsymbol{v}'_1,\boldsymbol{v}_2),\\ f(\boldsymbol{v}_1,a\boldsymbol{v}_2+b\boldsymbol{v}'_2) &= af(\boldsymbol{v}_1,\boldsymbol{v}_2)+bf(\boldsymbol{v}'_1,\boldsymbol{v}_2).\end{aligned}$$

It is easy to extend this definition to functions of more than two variables. Indeed, functions such as

$$f: V_1 \times V_2 \times \cdots \times V_n \to W \tag{5.24}$$

are called **multilinear functions**, more specifically n**-linear functions**, for which the defining relation is

$$f(\boldsymbol{v}_1,\cdots,a\boldsymbol{v}_i+b\boldsymbol{v}'_i,\cdots,\boldsymbol{v}_n) = af(\boldsymbol{v}_1,\cdots,\boldsymbol{v}_i,\cdots,\boldsymbol{v}_n)+bf(\boldsymbol{v}_1,\cdots,\boldsymbol{v}'_i,\cdots,\boldsymbol{v}_n). \tag{5.25}$$

An n-linear function can be multiplied by a scalar, and two n-linear functions can be added; in each case the result is an n-linear function. Thus, the set of n-linear functions given in (5.24) forms a vector space denoted by $\mathscr{L}(V_1 \times \cdots \times V_n,\, W)$.

---

[3]The notion of **direct product**, designated by $V \times W$, is completely different from the notion of **direct sum**, symbolized by $V+W$ (or the alternative notation $V \oplus W$ is sometime used). A vector space X is called the direct sum of its subspaces V and W, if for every $\boldsymbol{x} \in X$ there exists a unique expansion

$$\boldsymbol{x} = \boldsymbol{x}_1+\boldsymbol{x}_2,\ \boldsymbol{x}_1 \in V,\ \boldsymbol{x}_2 \in W.$$

The uniqueness of expansion means that

$$\boldsymbol{x} = \boldsymbol{y}_1+\boldsymbol{y}_2,\ \boldsymbol{y}_1 \in V,\ \boldsymbol{y}_2 \in W \quad \Rightarrow \quad \boldsymbol{x}_1 = \boldsymbol{y}_1 \text{ and } \boldsymbol{x}_2 = \boldsymbol{y}_2.$$

Every vector space having dimension greater than one can be represented by a direct sum of nonintersecting subspaces.

5.3.3 *Tensor product*

Suppose that $\tau^1 \in V_1^*$ and $\tau^2 \in V_2^*$; that is, τ^1 and τ^2 are linear real-valued functions on V_1 and V_2, respectively. Then we can form a bilinear real-valued function such as

$$\tau^1 \otimes \tau^2 : V_1 \times V_2 \to \mathbb{R}. \tag{5.26}$$

The function $\tau^1 \otimes \tau^2$ transforms an ordered pair of vectors, designated by $(\boldsymbol{v}_1, \boldsymbol{v}_2)$, to the product of two real numbers: $\tau^1(\boldsymbol{v}_1)$ and $\tau^2(\boldsymbol{v}_2)$. This transformation is written by

$$\tau^1 \otimes \tau^2(\boldsymbol{v}_1, \boldsymbol{v}_2) = \tau^1(\boldsymbol{v}_1)\tau^2(\boldsymbol{v}_2). \tag{5.27}$$

Note that the right side in (5.27) is just the product of two real numbers: $\tau^1(\boldsymbol{v}_1)$ and $\tau^2(\boldsymbol{v}_2)$. The bilinear function $\tau^1 \otimes \tau^2$ is called **tensor product** of τ^1 and τ^2. Clearly, since τ^1 and τ^2 are separately linear, $\tau^1 \otimes \tau^2$ is linear, too. Hence the set of tensor product $\tau^1 \otimes \tau^2$ forms a vector space $\mathscr{L}(V_1 \times V_2, \mathbb{R})$.

Recall that the vectors $\boldsymbol{v} \in V$ can be regarded as linear functions acting on V^*. In this context, we can also construct tensor products of two vectors. For example, let $\boldsymbol{v}_1 \in V_1$ and $\boldsymbol{v}_2 \in V_2$, and define the tensor product

$$\boldsymbol{v}_1 \otimes \boldsymbol{v}_2 : V_1^* \times V_2^* \to \mathbb{R} \tag{5.28}$$

by

$$\boldsymbol{v}_1 \otimes \boldsymbol{v}_2(\tau_1, \tau_2) = \boldsymbol{v}_1(\tau_1)\boldsymbol{v}_2(\tau_2) = \tau_1(\boldsymbol{v}_1)\tau_2(\boldsymbol{v}_2). \tag{5.29}$$

This shows that the tensor product $\boldsymbol{v}_1 \otimes \boldsymbol{v}_2$ can be considered as a bilinear function acting on $V_1^* \times V_2^*$, similar to that $\tau_1 \otimes \tau_2$ is a bilinear function on $V_1 \times V_2$. This means that the set of $\boldsymbol{v}_1 \otimes \boldsymbol{v}_2$ form a vector space $\mathscr{L}(V_1^* \times V_2^*, \mathbb{R})$.

Given a vector space V, furthermore, we can construct mixed type of tensor products such as

$$\boldsymbol{v} \otimes \tau : V^* \times V \to \mathbb{R} \tag{5.30}$$

given by

$$\boldsymbol{v} \otimes \tau(\phi, \boldsymbol{u}) = \boldsymbol{v}(\phi)\tau(\boldsymbol{u}) = \phi(\boldsymbol{v})\boldsymbol{u}(\tau), \tag{5.31}$$

where $\boldsymbol{u}, \boldsymbol{v} \in V$ and $\phi, \tau \in V^*$. In a straightforward extension, it is possible to develop tensor products of more than two linear functions or vectors such as

$$\boldsymbol{v}_1 \otimes \boldsymbol{v}_2 \otimes \cdots \boldsymbol{v}_r \otimes \tau^1 \otimes \tau^2 \otimes \cdots \tau^s, \tag{5.32}$$

which acts on the vector space

$$V^* \times V^* \times \cdots \times V^* \times V \times V \times \cdots \times V, \tag{5.33}$$

where V^* appears r times and V does s times. Similar to the previous cases, the set of tensor products (5.32) forms a vector space denoted by

$$\mathscr{L}\left[(V^*)^r \times V^s, \mathbb{R}\right], \tag{5.34}$$

where $(V^*)^r \times V^s$ is a direct product of V^* with r factors and those of V with s factors.

Keypoint:
Tensor product $(\boldsymbol{v}_1 \otimes \cdots \boldsymbol{v}_r \otimes \boldsymbol{\tau}^1 \otimes \cdots \boldsymbol{\tau}^s)$ is a multilinear function.
Direct product $(V^* \times \cdots \times V^* \times V \times \cdots \times V)$ is a vector space.

5.3.4 General definition of tensors

We finally reach the following generalized definition of tensors.

Definition (Tensor):
Let V be a vector space with dual space V^*. Then a **tensor of type** (r,s), denoted by $\boldsymbol{T}_s^{\ r}$, is a multilinear function

$$\boldsymbol{T}_s^{\ r} : (V^*)^r \times (V)^s \to \mathbb{R}. \tag{5.35}$$

The number r is called **contravariant degree** of the tensor, and s is called **covariant degree** of the tensor. ■

Definition (Tensor space):
The set of all tensors $\boldsymbol{T}_s^{\ r}$ for fixed r and s forms a vector space, called a **tensor space**, denoted by

$$\mathscr{T}_s^{\ r}(V) \equiv \mathscr{L}[(V^*)^r \times V^s, \mathbb{R}]. \tag{5.36}$$

As an example, let $\boldsymbol{v}_1, \cdots, \boldsymbol{v}_r \in V$ and $\boldsymbol{\tau}^1, \cdots, \boldsymbol{\tau}^s \in V^*$, and define the tensor product (i.e., multilinear function)

$$\boldsymbol{T}_s^{\ r} \equiv \boldsymbol{v}_1 \otimes \cdots \otimes \boldsymbol{v}_r \otimes \boldsymbol{\tau}^1 \otimes \cdots \otimes \boldsymbol{\tau}^s, \tag{5.37}$$

which yields for $\boldsymbol{\theta}^1, \cdots, \boldsymbol{\theta}^r \in V^*$ and $\boldsymbol{u}_1, \cdots, \boldsymbol{u}_s \in V$,

$$\begin{aligned} & \boldsymbol{v}_1 \otimes \cdots \otimes \boldsymbol{v}_r \otimes \boldsymbol{\tau}^1 \otimes \cdots \otimes \boldsymbol{\tau}^s(\boldsymbol{\theta}^1, \cdots, \boldsymbol{\theta}^r, \boldsymbol{u}_1, \cdots, \boldsymbol{u}_s) \\ = \; & \boldsymbol{v}_1(\boldsymbol{\theta}^1) \cdots \boldsymbol{v}_r(\boldsymbol{\theta}^r) \boldsymbol{\tau}^1(\boldsymbol{u}_1) \cdots \boldsymbol{\tau}^s(\boldsymbol{u}_s) \\ = \; & \prod_{i=1}^{r} \prod_{j=1}^{s} \boldsymbol{v}_i(\boldsymbol{\theta}^i) \boldsymbol{\tau}^j(\boldsymbol{u}_j). \end{aligned} \tag{5.38}$$

Observe that each $\boldsymbol{v}$ in the tensor product (5.38) requires an element $\theta \in V^*$ to produce a real number $\boldsymbol{\theta}$. This is why the number of factors of V^* in the direct product (5.37) equals the number of $\boldsymbol{v}$ in the tensor product (5.38).

In particular, a tensor of type $(0,0)$ is defined to be a **scalar**, so $\mathscr{T}_0^{\ 0}(V) = \mathbb{R}$. A tensor of type $(1,0)$ written by

$$\boldsymbol{T}_0^1 \equiv \boldsymbol{v} : V^* \to \mathbb{R}, \tag{5.39}$$

is called a **contravariant vector** [= called simply by a "vector" in elementary linear algebra]; one of type $(0,1)$ written by

$$T_1^0 \equiv \tau : V \to \mathbb{R}, \tag{5.40}$$

is called a **covariant vector** [= called simply by a "linear function" in elementary linear algebra]. More generally, a tensor of type $(r,0)$ is called a **contravariant tensor** of rank (or degree) r, and one of type $(0,s)$ is called a **covariant tensor** of rank (or degree) s.

We can form a tensor product of two tensors $T_s{}^r$ and U_ℓ^k such as

$$T_s{}^r \otimes U_\ell^k : (V^*)^{r+k} \times V^{s+\ell} \to \mathbb{R}, \tag{5.41}$$

which is a natural generalization of tensor products given in (5.27), (5.29) and (5.31). It is easy to prove that the tensor product is associative and distributive over tensor addition, but not commutative.

> **Keypoint:** A tensor $T_s{}^r$ is a linear function that maps collectively a set of s elements in V and r elements in V^* onto a real number.

5.4 Component of tensor

5.4.1 Basis of a tensor space

In physical applications of tensor calculus, it is necessary to choose a basis for the vector space V and one for its dual space V^* to represent the tensors by a set of numbers (i.e., components). The need for this process is analogous to the cases of elementary vector calculus, in which linear operators are often represented by arrays of numbers, that is, by matrices by referring a chosen basis of the space. A basis of our tensor space $\mathscr{T}_s{}^r(V) \equiv \mathscr{L}[(V^*)^r \times V^s, \mathbb{R}]$ is defined as follows.

Definition (Basis of a tensor space):
Let $\{e_i\}$ and $\{\varepsilon^j\}$ be a basis in V and V^*, respectively. Both V and V^* are N-dimensional vector spaces. Then, a **basis of the tensor space** $\mathscr{T}_s{}^r(V)$ is a set of all tensor products such as

$$e_{i_1} \otimes \cdots \otimes e_{i_r} \otimes \varepsilon^{j_1} \otimes \cdots \otimes \varepsilon^{j_s}. \tag{5.42}$$

Here, each of the $r+s$ indexes (i.e., $i_1, i_2, \cdots, i_r, j_1, j_2, \cdots, j_s$) takes integers ranging from 1 and N. ■

Note that for every factor in the basis of $\mathscr{T}_s{}^r(V)$ given by (5.42), there are N possibilities, wherein N is the dimension of V (and V^*). For instance, we have N choices for e_{i_1} in which $i_1 = 1, 2 \cdots, N$. Thus, the number of possible tensor products represented by (5.42) is N^{r+s}.

Definition (Components of a tensor):
The components of any tensor $\boldsymbol{A} \in \mathscr{T}_s^r(V)$ are the real numbers given by

$$A^{i_1 \cdots i_r}_{j_1 \cdots j_s} = \boldsymbol{A}\left(\varepsilon^{i_1}, \cdots, \varepsilon^{i_r}, \boldsymbol{e}_{j_1}, \cdots, \boldsymbol{e}_{j_s}\right). \quad \blacksquare \tag{5.43}$$

The above definition of the component of a tensor indicates the relation

$$\boldsymbol{A} = A^{i_1 \cdots i_r}_{j_1 \cdots j_s} \boldsymbol{e}_{i_1} \otimes \cdots \otimes \boldsymbol{e}_{i_r} \otimes \varepsilon^{j_1} \otimes \cdots \otimes \varepsilon^{j_s}. \tag{5.44}$$

Example 1:
A tensor space $\mathscr{T}_0^1(V)$ has a basis $\{\boldsymbol{e}_i\}$ so that an element (i.e., a contravariant vector) $\boldsymbol{v} \in \mathscr{T}_0^1(V)$ can be expanded by

$$\boldsymbol{v} = v^i \boldsymbol{e}_i. \tag{5.45}$$

Here the real numbers $v^i = \boldsymbol{v}(\varepsilon^i)$ are called the components of $\boldsymbol{v} : V \to \mathbb{R}$. $\blacksquare$

Example 2:
A tensor space $\mathscr{T}_1^0(V)$ has a basis $\{\varepsilon^j\}$ so that an element (i.e., a covariant vector) $\boldsymbol{\tau} \in \mathscr{T}_1^0(V)$ can be expanded by

$$\boldsymbol{\tau} = \tau_j \varepsilon^j. \tag{5.46}$$

Here the real numbers $\tau_j = \boldsymbol{\tau}(\boldsymbol{e}_j)$ are called the components of $\boldsymbol{\tau} : V^* \to \mathbb{R}$. $\blacksquare$

Example 3:
A tensor space $\mathscr{T}_1^2(V)$ has a basis $\{\boldsymbol{e}_i \otimes \boldsymbol{e}_j \otimes \varepsilon^k\}$ so that for any $\boldsymbol{A} \in \mathscr{T}_1^2(V)$ we have

$$\boldsymbol{A} = A^{ij}_k \boldsymbol{e}_i \otimes \boldsymbol{e}_j \otimes \varepsilon^k. \tag{5.47}$$

Here the real numbers

$$A^{ij}_k = \boldsymbol{A}\left(\varepsilon^i, \varepsilon^j, \boldsymbol{e}_k\right) \tag{5.48}$$

are the components of $\boldsymbol{A} : V \times V \times V^* \to \mathbb{R}$. $\blacksquare$

Keypoint: An indexed quantity, which is called by a "tensor" in physics and engineering, is a component of a tensor as mapping.

5.4.2 Transformation law of tensors

The components of a tensor depend on the basis in which they are described. If the basis is changed, the components change. The relation between components of a tensor in different bases is called **transformation law** for that particular tensor. Let us investigate this concept.

Given V, we can construct two different bases denoted by $\{\boldsymbol{e}_i\}$ and $\{\boldsymbol{e}'_i\}$. Similarly, we denote by $\{\varepsilon^j\}$ and $\{\varepsilon'^j\}$ two different bases of V^*. We can find appropriate transformation matrices $[R_i^j]$ and $[S_k^\ell]$ that satisfy the relations:

$$\boldsymbol{e}'_i = R_i^j \boldsymbol{e}_j \quad \text{and} \quad \varepsilon'^k = S_\ell^k \varepsilon^\ell. \tag{5.49}$$

In the case of a tensor $\boldsymbol{T}$ of type $(1,2)$, for instance, components of the tensor read as

$$\begin{aligned} T'^{i}_{\ jk} &= \boldsymbol{T}\left(\varepsilon'^{i}, \boldsymbol{e}'_{j}, \boldsymbol{e}'_{k}\right) = \boldsymbol{T}\left(S^{i}_{\ell}\varepsilon^{\ell}, R^{m}_{j}\boldsymbol{e}_{m}, R^{n}_{k}\boldsymbol{e}_{n}\right) \\ &= S^{i}_{\ell}R^{m}_{j}R^{n}_{k}\boldsymbol{T}\left(\varepsilon^{\ell}, \boldsymbol{e}_{m}, \boldsymbol{e}_{n}\right) \\ &= S^{i}_{\ell}R^{m}_{j}R^{n}_{k}T^{\ell}_{mn}. \end{aligned} \tag{5.50}$$

This is the transformation law of the components of the tensor $\boldsymbol{T}$ of type $(1,2)$.

Remember that in this coordinate-*dependent* treatment, a tensor of type $(1,2)$ was defined to be a collection of numbers, T^{m}_{np}, that transformed to another collection of numbers $T'^{i}_{\ jk}$ according to the rule in (5.50) when the basis was changed. In our current (i.e., coordinate-*free*) treatment of tensors, it is not necessary to introduce any basis to define tensors. A basis is to be introduced only when the components of a tensor are needed. The advantage of the coordinate-free approach is obvious, since a $(1,2)$ type tensor has 27 components in three dimensions and 64 components in four dimensions,[4] and all of these can be represented by the single symbol $\boldsymbol{T}$.

5.4.3 Natural isomorphism

We comment below an important property that is specific to components of tensors $\boldsymbol{A} \in \mathscr{T}_1^{\,1}(V)$. We have known that tensors $\boldsymbol{A} \in \mathscr{T}_1^{\,1}(V)$ are bilinear functions such as

$$\boldsymbol{A} : V^* \times V \to \mathbb{R}, \tag{5.51}$$

and their components A^{i}_{j} are defined by

$$A^{i}_{j} = \boldsymbol{A}(\varepsilon^{i}, \boldsymbol{e}_{j}), \tag{5.52}$$

where each ε^{i} and $\boldsymbol{e}_j$ is a basis of V^* and V, respectively. Now we consider the matrix

$$[A^{i}_{j}] = \begin{bmatrix} A^{1}_{1} & A^{1}_{2} & \cdots & A^{1}_{n} \\ A^{2}_{1} & & \cdots & \cdots \\ \cdots & & & \cdots \\ A^{n}_{1} & \cdots & & A^{n}_{n} \end{bmatrix}, \tag{5.53}$$

whose elements A^{i}_{j} are the same as given in (5.52). We shall see that (5.53) is the matrix representation of a linear operator A in terms of the basis $\{e_i\}$ of V, which associates a vector $\boldsymbol{v} \in V$ to another $\boldsymbol{u} \in V$, that is,

$$A : V \to V. \tag{5.54}$$

A formal statement on this issue is given below.

---

[4]Note that this argument does not downplay the role of components. In fact, when it comes to actual calculations, we are forced to choose a basis and manipulate components.

Theorem (Natural isomorphism):
For any vector space V, there is a one-to-one correspondence (called a **natural isomorphism**) between a tensor $\boldsymbol{A} \in \mathscr{T}_1^1(V)$ and a linear operator $A \in \mathscr{L}(V,V)$. ■

Proof:
We write the tensor $\boldsymbol{A} \in \mathscr{T}_1^1(V)$ as

$$\boldsymbol{A} = A^i_j \boldsymbol{e}_i \otimes \boldsymbol{\varepsilon}^j. \tag{5.55}$$

Given any $\boldsymbol{v} \in V$, we obtain

$$\begin{aligned} \boldsymbol{A}(\boldsymbol{v}) &= \left(A^i_j \boldsymbol{e}_i \otimes \boldsymbol{\varepsilon}^j\right)(\boldsymbol{v}) = A^i_j \boldsymbol{e}_i \left[\boldsymbol{\varepsilon}^j(\boldsymbol{v})\right] \\ &= A^i_j \boldsymbol{e}_i \left[\boldsymbol{\varepsilon}^j(v^k \boldsymbol{e}_k)\right] = A^i_j \boldsymbol{e}_i \left(v^k \delta^j_k\right) \\ &= A^i_j v^j \boldsymbol{e}_i. \end{aligned} \tag{5.56}$$

Observe that $A^i_j v^j$ in the last term are numbers, while $\boldsymbol{e}_i \in V$. Hence, the object $\boldsymbol{A}(\boldsymbol{v})$ is a linear combination of bases $\boldsymbol{e}_i$ for V, that is,

$$\boldsymbol{A}(\boldsymbol{v}) \in V. \tag{5.57}$$

Denoting $\boldsymbol{A}(\boldsymbol{v})$ in (5.56) by $\boldsymbol{u} = u^i \boldsymbol{e}_i$, it follows that

$$u^i = A^i_j v^j, \tag{5.58}$$

in which u^i, v^j are contravariant components of vectors $\boldsymbol{u}, \boldsymbol{v} \in V$, respectively, in terms of the basis $\{\boldsymbol{e}_i\}$. The result (5.58) is identified with a matrix equation such as

$$\begin{bmatrix} u^1 \\ u^2 \\ \cdots \\ u^n \end{bmatrix} = \begin{bmatrix} A^1_1 & A^1_2 & \cdots & A^1_n \\ A^2_1 & & & \cdots \\ \cdots & & & \\ A^n_1 & A^n_2 & \cdots & A^n_n \end{bmatrix} \begin{bmatrix} v^1 \\ v^2 \\ \cdots \\ v^n \end{bmatrix} \tag{5.59}$$

Therefore, we have seen that given $\boldsymbol{A} \in \mathscr{T}_1^1(V)$, its components form the matrix representation $[A^i_j]$ of a linear operator A that transforms a vector $\boldsymbol{v} \in V$ to another $\boldsymbol{u} \in V$.

In converse, for any given linear operator on V with a matrix representation $[A^i_j]$ in terms of a basis of V, there exists a tensor $\boldsymbol{A} \in \mathscr{T}_1^1(V)$. This suggests the one-to-one correspondence between the tensor space $\mathscr{T}_1^1(V)$ and the vector space $\mathscr{L}(V,V)$ comprising the linear mapping such as $f: V \to V$. ■

A parallel discussion goes as for linear operator on V^*. In fact, for any $\tau \in V^*$, we have

$$\begin{aligned} \boldsymbol{A}(\tau) &= \left(A^i_j \boldsymbol{e}_i \otimes \boldsymbol{\varepsilon}^j\right)(\tau) = A^i_j \left[\boldsymbol{e}_i(\tau)\right] \boldsymbol{\varepsilon}^j \\ &= A^i_j \left[\boldsymbol{e}_i \left(\tau_k \boldsymbol{\varepsilon}^k\right)\right] \boldsymbol{\varepsilon}^j = A^i_j \left(\tau_k \delta^k_i\right) \boldsymbol{\varepsilon}^j \\ &= A^i_j \tau_i \boldsymbol{\varepsilon}^j, \end{aligned}$$

which means that $\boldsymbol{A}(\tau)$ is a linear combination of bases ε^j for V^*, that is,

$$\boldsymbol{A}(\tau) \in V^*. \tag{5.60}$$

Denoting $\boldsymbol{A}(\tau)$ by $\theta = \theta_j \varepsilon^j$, we obtain

$$\theta_j = A^i_{\ j} \tau_i, \tag{5.61}$$

where θ_j and τ_i are (covariant) components of the *vectors* $\theta, \tau \in V^*$ in terms of the **basis** $\{\varepsilon^i\}$. Using the same matrix representation of $[A^i_{\ j}]$ as in (5.59), the result (5.61) can be rewritten by

$$[\theta_1, \cdots, \theta_n] = [\tau_1, \cdots, \tau_n] \begin{bmatrix} A^1_1 & A^1_2 & \cdots & A^1_n \\ A^2_1 & & & \cdots \\ \cdots & & & \\ A^n_1 & A^n_2 & \cdots & A^n_n \end{bmatrix}. \tag{5.62}$$

This describes a linear mapping from a vector $\tau \in V^*$ to another $\theta \in V^*$ through the linear operator with the matrix representation $[A^i_{\ j}]$.

Therefore, it is concluded that there is a natural isomorphism between the three vector spaces:

$$\mathscr{T}_1^{\ 1}(V) = \mathscr{L}(V^* \times V, \mathbb{R}), \quad \mathscr{L}(V,V), \quad \text{and}\ \mathscr{L}(V^*,V^*). \tag{5.63}$$

Due to the isomorphism, these three vector spaces can be treated as being the same.

Chapter 6

Lebesgue Integral

6.1 Motivation and merits

6.1.1 Merits of studying Lebesgue integral

We know that **Riemann integral** is the most standard and useful enough for applications in physics and engineering. Quite many functions we encounter in practice are continuous (piecewise, at least) so that they are integrable in the sense of Riemann procedure. In advanced subjects including functional analysis, however, we encounter a class of *highly irregular* functions or even pathological functions that are *everywhere* discontinuous, to which the concept of ordinary Riemann integral no longer applies to. In order to treat such functions, we need to employ another notion of the integral more flexible than that of the Riemann integral. In this chapter, we give a concise description as to what **Lebesgue integral** is. The Lebesgue integral not only overcome many difficulties inherent in the Riemann integral, but also provides a basic for constructing a specific class of **Hilbert spaces**, called L^p **spaces**.

One reason why Lebesgue integrals are so important is that they allow us to exchange the order of integration and other limiting procedures under very weak conditions. In fact, in the case of Riemann integrals, rather strict conditions are required for the following identities to be valid:

$$\lim_{n\to\infty}\int_{-\infty}^{\infty} f_n(x)dx = \int_{-\infty}^{\infty}\left[\lim_{n\to\infty} f_n(x)\right]dx, \tag{6.1}$$

and

$$\sum_{n=1}^{\infty}\int_{-\infty}^{\infty} f_n(x)dx = \int_{-\infty}^{\infty}\left[\sum_{n=1}^{\infty} f_n(x)\right]dx. \tag{6.2}$$

In a realm of Riemann integrals, the above equalities hold only if the integrands on the right side, i.e.,

$$\lim_{n\to\infty} f_n(x) \text{ and } \sum_{n=1}^{\infty} f_n(x), \tag{6.3}$$

are **uniformly convergent**[1] (see Appendix E.2 for the definition of uniform convergence). Such a restriction can be removed by using a Lebesgue integral since in the case of the latter, only **pointwise convergence** of the integrand is needed.[2] As a consequence, functions having discontinuity or divergence (or both of them) can be easily manipulated in integration.

6.1.2 Closer look to Riemann integral

It is pedagogical to review the nature of Riemann integral before discussing Lebesgue integrals. When defining the Riemann integral of a function $f(x)$ on an interval $I = [a,b]$, we divide the entire interval $[a,b]$ into small subintervals $\Delta x_k = [x_k, x_{k+1}]$ such that

$$a = x_1 < x_2 < \cdots < x_{n+1} = b. \tag{6.5}$$

The finite set $\{x_i\}$ of numbers is called a **partition** P of the interval I. Using this notation P, let us define the sum such as

$$S_P(f) = \sum_{k=1}^{n} M_k(x_{k+1} - x_k), \quad s_P(f) = \sum_{k=1}^{n} m_k(x_{k+1} - x_k), \tag{6.6}$$

where M_k and m_k are the **supremum** and **infimum** of $f(x)$ on the interval $\Delta x_k = [x_k, x_{k+1}]$, respectively, given by

$$M_k = \sup_{x\in\Delta x_k} f(x), \quad m_k = \inf_{x\in\Delta x_k} f(x). \tag{6.7}$$

Evidently, the relation $S_P(f) \geq s_P(f)$ holds if the function $f(x)$ is bounded on the interval $I = [a,b]$. We take the **limit inferior** (or **limit superior**) of the sums such as

$$S(f) = \liminf_{n\to\infty} S_P, \quad s(f) = \limsup_{n\to\infty} s_P, \tag{6.8}$$

---

[1]A counterexample is given by $f_n(x) = 2n^2 x e^{-n^2x^2}$ in the interval $[0,1]$. The $f_n(x)$ is Riemann integrable for all n; as well, its limit $f(x) \equiv \lim_{n\to\infty} f_n(x) = 0$ is a constant function and thus integrable, too. But the equality (6.1) fails because

$$\lim_{n\to\infty}\int_{-\infty}^{\infty} f_n(x)dx = \lim_{n\to\infty}\left(1 - e^{-n^2}\right) = 1 \text{ and } \int_{-\infty}^{\infty}\left[\lim_{n\to\infty} f_n(x)\right]dx = \int_0^1 0dx = 0. \tag{6.4}$$

The disagreement stems from that the convergence of $f_n(x)$ to $f(x)[\equiv 0]$ is not uniform but pointwise.

[2]Exactly speaking, certain auxiliary conditions must be imposed on the sequence of functions, in addition to the pointwise convergence. See section 6.4.2 for details.

where all possible choices of partition P are taken into account. The $S(f)$ and $s(f)$ are called the **upper** and **lower Riemann-Darboux integrals** of f over I, respectively. If the relation holds such as

$$S(f) = s(f) = A, \tag{6.9}$$

the common value A is called the **Riemann integral** and the function $f(x)$ is called **Riemann integrable** such that

$$A = \int_a^b f(x)dx. \tag{6.10}$$

Otherwise, $f(x)$ is not integrable in the sense of Riemann procedure; the notation of (6.10) does not make sense there.

Keypoint: Definition of Riemann integral is based on the equality: $S(f) = s(f)$.

We note without proof that continuity (piecewise at least) of an integrand $f(x)$ ensure the existence of the Riemann integral of $f(x)$. Contrariwise, when the function $f(x)$ exhibits too many points of discontinuity, the above definition is of no use to form the integral. An illustrative example is given below.

Example:
Consider an enumeration $\{z_n\}$ $(n = 1, 2, \cdots)$ of the rational numbers between 0 and 1, and let

$$f(x) = \begin{cases} 1 & (x = z_1, z_2, \cdots, z_n), \\ 0 & \text{otherwise.} \end{cases} \tag{6.11}$$

That is, the function $f(x)$ has the value unity if x is rational and the value zero if x is irrational. In any subdivision of the interval $\Delta x_k \subset [0, 1]$,

$$m_k = 0, \quad M_k = 1, \tag{6.12}$$

and

$$s_P = 0, \quad S_P = 1. \tag{6.13}$$

Therefore, the upper and lower Darboux integrals are 1 and 0, respectively, whence $f(x)$ has no Riemann integral. ■

Keypoint: Riemann integration is of no use for highly discontinuous functions.

The shortcoming of Riemann's procedure demonstrated above can be successfully overcome by employing Lebesgue's procedure.

6.2 Measure theory

6.2.1 Measure

The Lebesgue procedure requires a systematic way of assigning a **measure** $\mu(X_i)$ to each subset of points X_i. The measure for a subset of points is a generalization of the concept of the length, area, and volume. Intuitively, it follows that the length of an interval $[a,b]$ is $b-a$. Similarly, if we have two disjoint intervals $[a_1,b_1]$ and $[a_2,b_2]$, it is natural to interpret the length of the set consisting of these two intervals as the sum $(b_1-a_1)+(b_2-a_2)$. However, it is not obvious what is the "length" of a set of points of rational (or irrational) numbers on the line. This context requires a rigorous mathematical definition of a measure of a point set. It is given as follows.

Definition (Measure):
A measure $\mu(X)$ defined on a set of points X is a function with the properties:

1. If the set X is empty or consists of a single point, $\mu(X)=0$; otherwise, $\mu(X)>0$.
2. The measure of the sum of two non-overlapping sets is equal to the sum of the measures of these sets expressed by

$$\mu(X_1+X_2)=\mu(X_1)+\mu(X_2), \quad \text{for } X_1\cap X_2=0. \quad \blacksquare \tag{6.14}$$

In the above statement, X_1+X_2 denotes the set containing both elements of X_1 and X_2, wherein each element is counted only once. If X_1 and X_2 overlap, (6.14) is replaced by

$$\mu(X_1+X_2)=\mu(X_1)+\mu(X_2)-\mu(X_1\cap X_2) \tag{6.15}$$

in order to count only once the points common to X_1 and X_2.

Keypoint: Measure $\mu(X)$ is a length of the set X.

6.2.2 Lebesgue measure

The procedure of the Lebesgue integral essentially reduces to that of *finding a measure for sets of arguments*. Particularly, if a set consists of too many points of discontinuity, we need some ideas to define its measure; the resulting measure is called the **Lebesgue measure**. In this subsection, we explain how to construct the Lebesgue measure of a point set.

As a simple example, let us consider a finite interval $[a,b]$ of length L. This can be decomposed into two sets. The one is a set X consisting of a part of points $x\in[a,b]$, and the other is its **complementary set** X' consisting of all points $x\in[a,b]$ which do not belong to X. A schematic view of X and X' is given in Fig. 6.1. Each of X and X' may be a set of several continuous line segments or a set of **isolated points**.

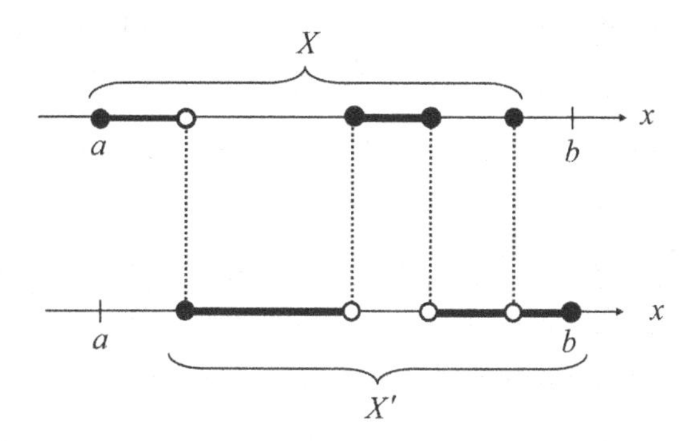

Figure 6.1: A set X and its complementary set X'.

We would like to evaluate the Lebesgue measure of X. To this aim, we cover the set of points X by non-overlapping intervals $\Lambda_i \subset [a,b]$ such as

$$X \subset (\Lambda_1 + \Lambda_2 + \cdots). \tag{6.16}$$

If we denote the length of Λ_k by ℓ_k, the sum of ℓ_k must satisfy the inequality:

$$0 \leq \sum_k \ell_k \leq L. \tag{6.17}$$

In particular, the smallest value of the sum $\sum_i \ell_i$ is referred to as the **outer measure** of X, denoted by

$$\mu_{\text{out}}(X) = \inf\left(\sum_k \ell_k\right). \tag{6.18}$$

In the same manner, we can find intervals $\Lambda_k{}' \subset [a,b]$ of lengths $\ell_1', \ell_2', \cdots$ which cover the complementary set X' such that

$$X' \subset (\Lambda_1' + \Lambda_2' + \cdots), \quad 0 \leq \sum_k \ell_k' \leq L. \tag{6.19}$$

Here we define another kind of measure denoted by

$$\mu_{\text{in}}(X) \equiv L - \mu_{\text{out}}(X') \ = L - \inf\left(\sum_k \ell_k'\right), \tag{6.20}$$

which is called the **inner measure** of X. Notice that the inner measure of X is defined by the outer measure of X', not of X. It is straightforward to prove that the inequality

$$0 \leq \mu_{\text{in}}(X) \leq \mu_{\text{out}}(X). \tag{6.21}$$

Specifically, if we obtain

$$\mu_{\text{in}}(X) = \mu_{\text{out}}(X), \tag{6.22}$$

this is called the **Lebesgue measure** of the point set X, denoted by $\mu(X)$. Clearly, when X contains all the points of $[a,b]$, the smallest interval that covers $[a,b]$ is $[a,b]$ itself, and thus $\mu(X) = L$.

Our results are summarized by:

Definition (Lebesgue measure):
A set of points X is said to be **measurable** with the **Lebesgue measure** $\mu(X)$ if and only if $\mu_{\text{in}}(X) = \mu_{\text{out}}(X) \equiv \mu(X)$. ■

6.2.3 *Countable set*

It is easy to evaluate Lebesgue measure of a continuous line segment with finite length L; it is equal to L. Evaluation is easy, too, for an isolated point; its Lebesgue measure equals to zero. But it becomes nontrivial if an infinite number of isolated points is given. For instance, consider a set of all rational numbers within the interval $[0,1]$ and try to evaluate its Lebesgue measure. Certainly, each point consisting of the set has zero Lebesgue measure. Nevertheless, the set is composed of infinitely many points; by putting them on a real number axis, therefore, we can construct an apparently continuous line segment whose Lebesgue measure seems to have a nonzero value. To resolve the problem, we need to learn the way how to *count* the number of infinitely many points.

In mathematics, sets (finite or infinite) are categorized to either **countable** or **uncountable** sets. A rigorous definition of countable sets is given herewith:

Definition (Countable set):
A finite or infinite set X is **countable**[3] if and only if it is possible to establish a reciprocal one-to-one correspondence between its elements and the elements of a set of real integers. ■

It is known that every finite set is countable, and every subset of a countable set is also countable. Any countable set is associated with a specific number, called the **cardinal number**, as defined below.

Definition (Cardinal number):
Two sets X_1 and X_2 are said to have the same **cardinal number** if and only if there exists a reciprocal one-to-one correspondence between their respective elements. ■

The notion of cardinal number gives a criterion for the finiteness/infiniteness of sets as shown below.

[3] Or it is called **enumerable**.

Definition (Finite and Infinite sets):
A set X is called an **infinite set** if it has the same cardinal number as one of its subsets; otherwise, X is called a **finite set**. ■

	Countable	Uncountable
Finite	$\{1,2,\ldots,100\}$	╳
Infinite	$\mathbb{N}, \mathbb{Z}, \mathbb{Q}$	$\mathbb{R}, \mathbb{C}$

Figure 6.2: Table categorizing number sets into countable/uncountable and finite/infinite ones.

It should be stressed that an infinite set may or may not be countable. When a given infinite set is countable, then its cardinal number is denoted by $\aleph_0$, which is the same as the cardinal number of the set of the positive real integers. As well, the cardinal number of every non-countable set is denoted by $2^{\aleph_0}$, which is identified with the cardinal number of the set of all real numbers (or the set of points of a continuous line). Cardinal numbers of infinite sets, $\aleph_0$ and $2^{\aleph_0}$, are called **transfinite numbers**.

The most important property of countable sets in view of the measure theory is given below:

Theorem:
Any countable sets (finite or infinite) has a Lebesgue measure of zero, namely, **null measure**. ■

Example 1:
We show below that the set of all rational numbers in the interval $[0,1]$ has the Lebesgue measure equal to zero.

Denote by X' the set of irrational numbers that is complementary to X, and by I the entire interval $[0,1]$. Since $\mu(I) = 1$, the outer measure of X' reads

$$\mu_{\text{out}}(X') = \mu_{\text{out}}(I - X) = \mu_{\text{out}}(I) - \mu_{\text{out}}(X) = 1 - \mu_{\text{out}}(X). \tag{6.23}$$

By definition, the inner measure of X is given by

$$\mu_{\text{in}}(X) = \mu_{\text{in}}(I) - \mu_{\text{out}}(X') = 1 - [1 - \mu_{\text{out}}(X)] = \mu_{\text{out}}(X). \tag{6.24}$$

The last equality asserts that the set X is measurable. The remained task is to evaluate the value of $\mu(X) = 0$.

Let x_k $(k = 1, 2, \cdots, n, \cdots)$ denote the points of rational numbers in the interval I. We cover each point $x_1, x_2, \cdots, x_n, \cdots$ by an open interval of length

$\frac{\varepsilon}{2}, \frac{\varepsilon}{2^2}, \cdots, \frac{\varepsilon}{2^n}, \cdots$, respectively, where ε is an arbitrary positive number. Since these intervals may overlap, the entire set can be covered by an open set of measure not greater than

$$\sum_{n=1}^{\infty} \frac{\varepsilon}{2^n} = \frac{\varepsilon}{2(1-\frac{1}{2})} = \varepsilon. \tag{6.25}$$

Since ε can be made arbitrarily small, we find that $\mu_{\text{out}}(X) = 0$. Hence, we immediately have $\mu(X) = 0$. ■

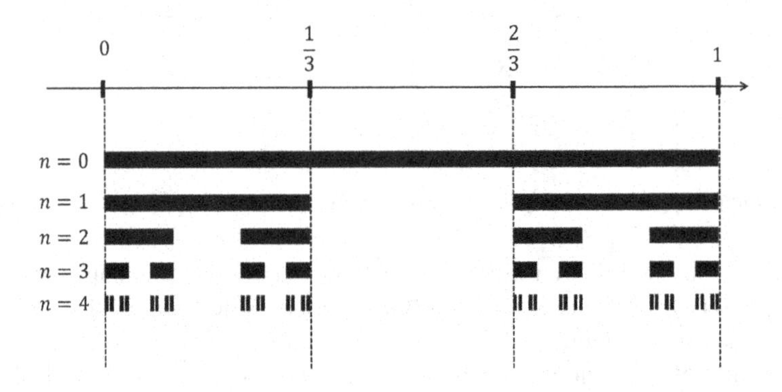

Figure 6.3: Diagram of constructing the Cantor set.

Example 2:
Another well-known example of the set of measure zero is the **Cantor set**. The Cantor set is an infinite set of discrete points constructed by:

i) Delete from the closed interval $[0,1]$ the open interval $(1/3, 2/3)$ which forms its middle third;
ii) Delete again from each of the remaining intervals $[0,1/3]$ and $[2/3,1]$ its middle third;
iii) Continue indefinitely this process of deleting the middle thirds to obtain the point set on the line that remains after all these open intervals.
Observe that at the kth step, we have thrown out 2^{k-1} adjacent intervals of length $1/3^k$. Thus the sum of the lengths of the intervals removed is equal to

$$\frac{1}{3} + \frac{2}{9} + \frac{4}{27} + \cdots + \frac{2^{n-1}}{3^n} + \cdots = \lim_{n\to\infty} \frac{\frac{1}{3}[1-(\frac{2}{3})^n]}{1-\frac{2}{3}} = 1. \tag{6.26}$$

This is just the measure of the open set P' that is the **complement** of P. Therefore, the Cantor set P itself has null measure

$$\mu(P) = 1 - \mu(P') = 1 - 1 = 0. \quad ■ \tag{6.27}$$

6.3 Lebesgue integral

6.3.1 What is Lebesgue integral?

We are in a position to define what the **Lebesgue integral** is. Let the function $f(x)$ be defined on a set X which is bounded such as

$$0 \le f_{\min} \le f(x) \le f_{\max}. \tag{6.28}$$

We partition the ordinate axis by the sequence $\{f_k\}$ $(1 \le k \le n)$ so that $f_1 = f_{\min}$ and $f_n = f_{\max}$. Due to the one-to-one correspondence between x and $f(x)$, there should exist sets X_i of values x such that

$$f_k \le f(x) < f_{k+1} \quad \text{for } x \in X_k, \quad (1 \le k \le n-1), \tag{6.29}$$

as well as a set X_n of values x such that $f(x) = f_n$. Each set X_k assumes a measure $\mu(X_k)$. Thus we form the sum of products $f_k \cdot \mu(X_k)$ of all possible values of f, called **Lebesgue sum**, such as

$$\sum_{k=1}^{n} f_k \cdot \mu(X_k). \tag{6.30}$$

If the sum (6.30) converges to a finite value when taking the limit $n \to \infty$ such that

$$\max |f_k - f_{k+1}| \to 0, \tag{6.31}$$

then the limiting value of the sum is called the **Lebesgue integral** of $f(x)$ over the set X.

The formal definition of the Lebesgue integral is given below.

Definition (Lebesgue integral):
Let $f(x)$ be a non-negative function defined on a **measurable set** X, and divide X into a finite number of subsets such as

$$X = X_1 + X_2 + \cdots + X_n. \tag{6.32}$$

Let $f_k = \inf_{x \in X_k} f(x)$ to form the sum

$$\sum_{k=1}^{n} f_k \mu(X_k). \tag{6.33}$$

Then the **Lebesgue integral** of $f(x)$ on X is defined by

$$\int_X f d\mu \equiv \lim_{\max|f_k - f_{k-1}| \to 0} \left[\sum_{k=1}^{n} f_k \mu(X_k) \right], \tag{6.34}$$

where all possible choices of partition (6.32) is considered. ■

Figure 6.4 gives a schematic illustration of the Lebesgue procedure. Obviously, the value of the Lebesgue sum (6.30) depends on our choice of partition; if we take an alternative partition instead of (6.32), the value of the sum also changes. Among the infinite variety of choices, such partition that makes the sum (6.32) be the maximum gives the Lebesgue integral of $f(x)$. That a function is **Lebesgue integrable** means that the superior of the sum (6.33) is determined independently of our choice of the partition of the x axis.

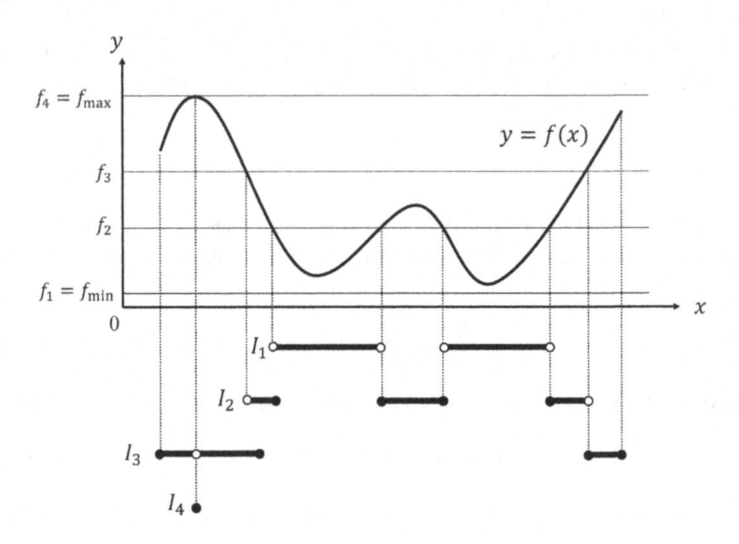

Figure 6.4: An illustration of the Lebesgue procedure.

6.3.2 *Riemann vs. Lebesgue*

Here is given a comparison between the definition of Riemann and Lebesgue integrals, which provides us a better understanding of the significance of the latter. In the language of measure, the **Riemann integral** of a function $f(x)$ defined on the set X is obtained by dividing X into non-overlapping subsets X_i as

$$X = X_1 + X_2 + \cdots + X_n, \quad X_i \cap X_j = 0, \text{ for any } i, j, \tag{6.35}$$

followed by setting the **Riemann sum** such as

$$\sum_{k=1}^{n} f(\xi_k)\mu(X_k). \tag{6.36}$$

Here, the measure $\mu(X_k)$ is identified with the length of the subset X_k, and ξ_k assumes any point that belongs to X_k. We next increase the number of subsets $n \to \infty$ such that

$$\mu(X_k) \to 0, \quad \text{for any } X_k. \tag{6.37}$$

Suppose that under the conditions, the limit of the sum (6.36),

$$\lim_{n\to\infty} \sum_{k=1}^{n} f(\xi_k)\mu(X_k), \tag{6.38}$$

exists and the limiting value is independent of the subdivision process. If this is true, the limit (6.38) is called the **Riemann integral** of $f(x)$ over X. Obviously, the Riemann integral can be defined under the condition that all values of $f(x)$ defined over X_k tend to a common limit as $\mu(X_k) \to 0$. Such a requirement excludes the possibility of defining the Riemann integral for functions having too many points of discontinuity.

Although the Lebesgue sum given in (6.30) is apparently similar to the Riemann sum given in (6.36), they are intrinsically different from each other. In the Riemann sum (6.36), $f(\xi_i)$ is the value of $f(x)$ at *arbitrary* point $\xi_i \in X_i$; thereby the value of ξ_i is allowed to vary within each subset, which causes an indefiniteness of the value of $f(\xi_i)$ within each subset. On the other hand, in the Lebesgue sum (6.30), the value of f_i corresponding to each subset X_i has a definite value. Therefore, for the existence of the Lebesgue integral, we no longer need the local smoothness of $f(x)$. As a result, the conditions imposed on the integrated function become very *weak* compared with the case of the Riemann integral.

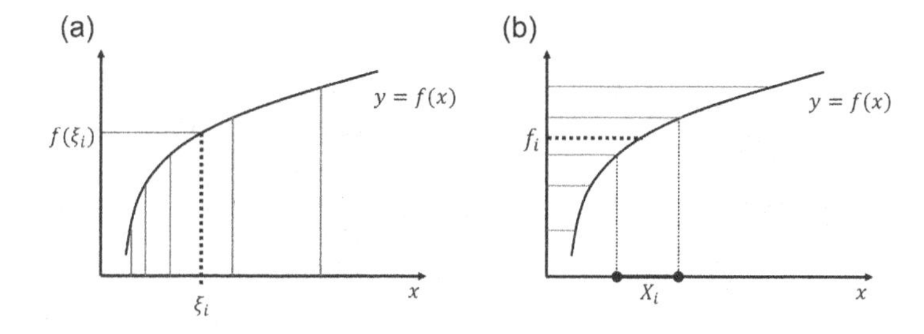

Figure 6.5: (a) Partition of the area below the curve $y = f(x)$ in the longitudinal direction. (b) That in the horizontal direction.

> **Keypoint:** Look at Fig. 6.5. The Riemann sum is based on the *longitudinal* partition; The Lebesgue sum is based on the *horizontal* partition.

6.3.3 Property of Lebesgue integral

Several properties of the Lebesgue integral are listed below without proof.

1. If $\mu(X) = 0$, then $\int_X f(x)dx = 0$.

2. If the integral $\int_X f(x)dx$ is finite, then the subset of X defined by

$$X' = \{x \mid f(x) = \pm\infty\} \tag{6.39}$$

has zero measure. This means that in order for the integral to converge, the measure of a set of points x at which $f(x)$ diverges should be necessarily zero.

3. Suppose $\int_X f(x)dx$ be finite and $X' \subset X$. If we make $\mu(X') \to 0$, then

$$\int_{X'} f d\mu \to 0. \tag{6.40}$$

4. When $f(x)$ on X takes both positive and negative values, its Lebesgue integral is defined by

$$\int_X f d\mu = \int_X f^+ d\mu + \int_X f^- d\mu, \tag{6.41}$$

and[4]

$$\int_X |f| d\mu = \int_X f^+ d\mu - \int_X f^- d\mu, \tag{6.42}$$

where

$$f^+(x) = \begin{cases} f(x) & \text{for } x \text{ at which } f(x) \geq 0, \\ 0 & \text{for } x \text{ at which } f(x) < 0, \end{cases} \tag{6.43}$$

and

$$f^-(x) = \begin{cases} 0 & \text{for } x \text{ at which } f(x) \geq 0, \\ -f(x) & \text{for } x \text{ at which } f(x) < 0. \end{cases} \tag{6.44}$$

6.3.4 Concept of "almost everywhere"

We have observed that sets of measure zero have no contribution to the Lebesgue integrals. This fact provides a concept of an **equality almost everywhere** for measurable functions. This concept plays an important role in developing the theory of function analysis.

Definition (Equality almost everywhere):
Two functions $f(x)$ and $g(x)$ defined on the same set X are said to be **equal almost everywhere** with respect to a measure $\mu(X)$ if

$$\mu\{x \in X;\ f(x) \neq g(x)\} = 0. \quad \blacksquare \tag{6.45}$$

[4]The definition (6.42) is justified except when two integrals in the right side diverges.

Example:
Consider again the function given by (6.11);

$$f(x) = \begin{cases} 1 & \text{at } x \text{ being a rational number within } [0,1, \\ 0 & \text{otherwise.} \end{cases} \tag{6.46}$$

Since every isolated point has null measure, $f(x)$ is equal to zero almost everywhere. ■

We will extend this terminology to other circumstances as well. In general, a property is said to hold **almost everywhere** on X if it holds at all points of X except on a set of measure zero. Since the behavior of functions on sets of measure zero is often unimportant, it is natural to introduce the following generalization of the ordinary notion of the convergence of a sequence of functions:

Definition (Convergence almost everywhere):
Consider a sequence of functions $\{f_n(x)\}$ defined on a set X. The sequence is said to **converge almost everywhere** to $f(x)$, if

$$\lim_{n\to\infty} f_n(x) = f(x) \tag{6.47}$$

holds for all $x \in X$ except at points of measure zero. ■

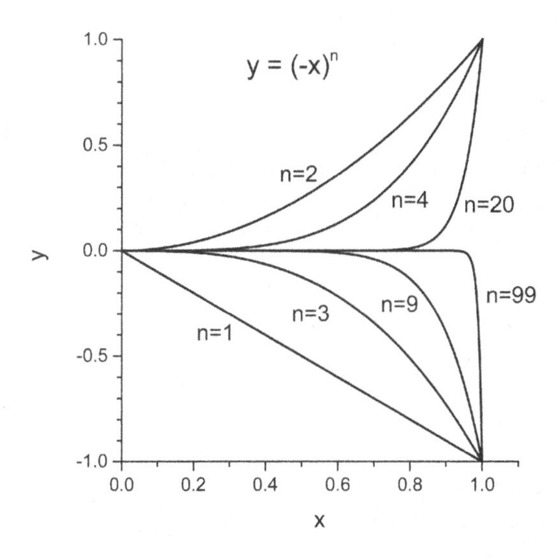

Figure 6.6: Plots of the function $y = (-x)^n$ for various n.

Example:
Consider the sequence of functions $\{f_n(x)\}$ defined by

$$f_n(x) = (-x)^n \quad \text{with } 0 \le x \le 1. \tag{6.48}$$

The sequence converges almost everywhere to the function $f(x) \equiv 0$, in fact everywhere except at the point $x = 1$. ■

6.4 Lebesgue convergence theorem

6.4.1 Monotone convergence theorem

In section 6.1.1, it was commented that in cases of Lebesgue integration, the order of integration and limiting procedure becomes exchangable under looser conditions than the case of Riemann integration. We now have a closer look to the issue, namely, what conditions are required to make the equality

$$\lim_{n\to\infty} \int_X f_n d\mu = \int_X f d\mu \tag{6.49}$$

be valid under the Lebesgue procedure.[5] This problem can be clarified by referring two important theorems concerning the convergence property of the Lebesgue integral; these are called **monotone convergence theorem** and **dominated convergence theorem**.[6] We will observe that, due to the two convergence theorems, the Lebesgue theory offers a considerable improvement over the Riemann theory with regard to convergence properties.

In what follows, we assume that X be a set of real numbers, and $\{f_n\}$ be a sequence of functions defined on X.

Monotone convergence theorem:

Consider a sequence $\{f_n\}$ that satisfies $0 \le f_n \le f_{n+1}$ for all $n \ge 1$ in X. If it converges pointwisely to f that is Lebesgue integrable in X, we have

$$\lim_{n\to\infty} \int_X f_n d\mu = \int_X \left[\lim_{n\to\infty} f_n\right] d\mu = \int_X f d\mu. \quad ■ \tag{6.52}$$

The monotone convergence theorem states that, in the case of Lebesgue integrals, only the **pointwise convergence** of $f_n(x)$ to $f(x)$ is required to reverse the order of limit and integration. This condition is much weaker than the case of Riemann integrals, in which the **uniform convergence** of $f_n(x)$ to $f(x)$ is necessary.

---

[5]The equality (6.49) sometimes fails even if f_n for all n and $\lim_{n\to\infty} f_n$ are both Lebesgue integrable. This fact follows from the example in which

$$f_n(x) = \begin{cases} 0 & \text{at } x = 0, \\ n & \text{for } 0 < x < 1/n, \\ 0 & \text{for } 1/n \le x \le 1. \end{cases} \tag{6.50}$$

The limit of the sequence $\{f_n\}$ reads as $f = \lim_{n\to\infty} f_n \equiv 0$ for all $X =\in [0,1]$, thus $\int_X f d\mu = 0$. On the other hand, we have

$$\int_X f_n d\mu = \int_0^{1/n} n dx + \int_{1/n}^1 0 dx = n \cdot \frac{1}{n} = 1. \tag{6.51}$$

This counterexample indicates the need of additional conditions in order for the equality (6.49) to hold. The conditions are demonstrated in the two theorems stated below.

[6]Both theorems fail if we restrict our attention to Riemann integrable functions.

6.4.2 *Dominated convergence theorem*

In the previous argument, we have learned that the order of limit and integration can be reversed when considering monotonically increasing sequences of functions. In practice, however, the requirement in the monotone convergence theorem—i.e., the sequence $\{f_n(x)\}$ should be monotonically increasing—is sometimes very inconvenient. In the following discussion, we examine the same issue for more general sequences of functions, i.e., non-monotone sequences satisfying some looser conditions and their limit passage.

Dominated convergence theorem:
Let $\{f_n\}$ be a sequence of functions defined **almost everywhere** on X such that:
1. $\lim_{n\to\infty} f_n(x) = f(x)$ exists, and
2. There exists a Lebesgue integrable function g on X such that $|f_n| \leq g$ for all $n \geq 1$.
Then, we have

$$\lim_{n\to\infty} \int_X f_n d\mu = \int_X f d\mu. \quad \blacksquare \tag{6.53}$$

Plainly speaking, the dominated convergence theorem requires that f_n for all n should be bounded from above **almost everywhere**. This condition is clearly looser than that imposed in the monotone convergence theorem. Hence, the monotone convergence theorem can be regarded as a special case of the dominated convergence theorem.

It is to be noted that every Riemann integrable function is Lebesgue integrable. Hence, the dominated convergence theorem can apply to Riemann integrable functions, too. To observe the efficiency of the theorem to practical calculation of Riemann integrals, take a look at the following example.

$$I = \lim_{n\to\infty} \int_0^n \left(1 + \frac{x}{n}\right)^n e^{-2x} dx. \tag{6.54}$$

It is possible to evaluate I through the partial integration technique; however, rather lengthy calculations are required as shown below.

$$\int_0^n \left(1 + \frac{x}{n}\right)^n e^{-2x} dx \tag{6.55}$$

$$= \frac{1}{2}\left(1 - 2^n e^{-2n}\right) + \frac{1}{2}\int_0^n \left(1 + \frac{x}{n}\right)^{n-1} e^{-2x} dx \tag{6.56}$$

$$= \frac{1}{2}\left(1 - 2^n e^{-2n}\right) + \frac{1}{2}\left(1 - 2^{n-1} e^{-2n}\right) + \frac{1}{2}\cdot\frac{n-1}{n}\left(1 - 2^{n-2} e^{-2n}\right) + \cdots \tag{6.57}$$

$$\to 1 \quad (n \to \infty). \tag{6.58}$$

Using the dominated convergence theorem, the result can be obtained more easily. We first note that

$$\left(1 + \frac{x}{n}\right)^n e^{-2x} \leq e^{-x} \quad \text{for } x > 0 \text{ and } n \geq 1, \tag{6.59}$$

and

$$\lim_{n\to\infty}\left(1+\frac{x}{n}\right)^n e^{-2x} = e^{-x} \quad \text{for } x > 0. \tag{6.60}$$

Since e^{-x} is Riemann integrable in the interval $[0,\infty)$, we can exchange the order of integration and limiting procedure in (6.54) to obtain

$$I = \int_0^\infty e^{-x} dx = 1. \tag{6.61}$$

The applicability of the theorem relies on the relation (6.59), which indicates that the integrand $\left(1+\frac{x}{n}\right)^n e^{-2x}$ is dominated by e^{-x} that is integrable on $0 < x < \infty$. The existence of such a dominant and integrable function is dispensable for applying the dominated convergence theorem. Two counterexamples given below will make clear the importance of the existence of dominant functions.

Example 1:
Consider a sequence of functions f_n defined on the point set $X = (0,\pi]$. Each f_n is given by

$$f_n(x) = \begin{cases} n \sin nx & \text{for } 0 < x < \dfrac{\pi}{n}, \\ 0 & \text{for } \dfrac{\pi}{n} \le x \le \pi. \end{cases} \tag{6.62}$$

As depicted in Fig. 6.7, f_n grows in height and shrinks in width with increasing n. Eventually at the limit $n \to \infty$, it converges to zero pointwisely at $x \in X$. On the other hand, we have

$$\lim_X f_n d\mu = \int_0^\pi f_n(x) dx = 2 \tag{6.63}$$

for any n. It is thus concluded that

$$\lim_{n\to\infty} \int_X f_n d\mu \neq \int_X \lim_{n\to\infty} f_n d\mu. \tag{6.64}$$

Namely, the dominated convergence theorem does not apply to this case, because of the loss of non-negative function g on X that satisfies $|f_n| < g$ for all n.

Example 2:
Consider the interval $X = [0,1]$ on which the following function f_n is defined.

$$f_n(x) = \begin{cases} n^\alpha & \text{for } 0 \le x \le \dfrac{1}{n}, \\ 0 & \text{for } \dfrac{1}{n} < x \le 1. \end{cases} \tag{6.65}$$

The limit of the sequence $\{f_n\}$ reads as

$$f(x) \equiv \lim_{n\to\infty} f_n(x) = 0 \quad \text{at } x \neq 0. \tag{6.66}$$

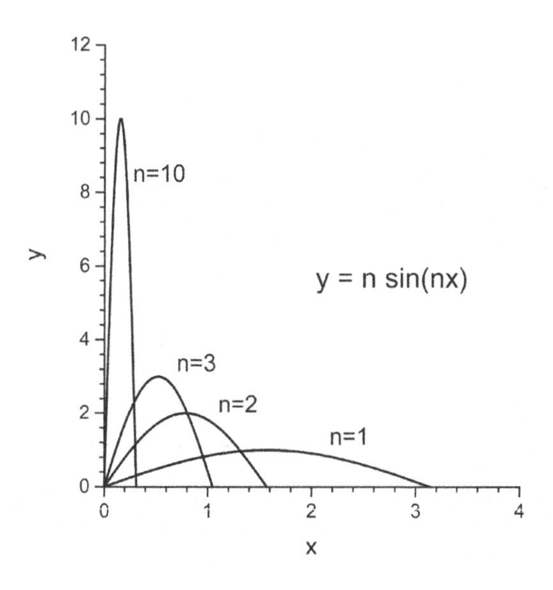

Figure 6.7: Plots of the function $y = n\sin(nx)$ for various n.

Namely, $f(x) \equiv 0$ **almost everywhere** except at the point $x = 0$ that has null measure. We thus obtain

$$\int_0^1 \lim_{n\to\infty} f_n(x)dx = \int_0^1 f(x)dx = \int_0^1 0dx = 0. \tag{6.67}$$

On the other hand,

$$\int_0^1 f_n(x)dx = n^{\alpha-1}, \tag{6.68}$$

and thus

$$\lim_{n\to\infty} \int_0^1 f_n(x)dx = \begin{cases} 0 & (\alpha < 1), \\ 1 & (\alpha = 1), \\ \infty & (\alpha > 1). \end{cases} \tag{6.69}$$

Therefore, the dominated convergence theorem fails when $\alpha \geq 1$. ■

6.4.3 *Remarks on the dominated convergence theorem*

It warrants caution that the existence of dominant function g, which satisfies $|f_n| \leq g$ for all n, is a *necessary* condition for the dominated convergence theorem. Hence, the conclusion of the theorem [i.e., that given by (6.53)] may and may not be true even when such function g cannot be found.

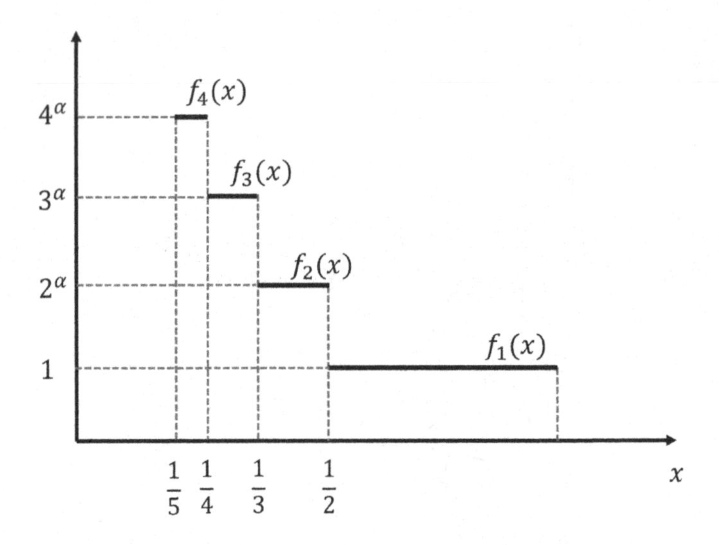

Figure 6.8: Plots of the function $y = f_n(x)$ given by (6.70).

This delicate issue is made clear by the following example. Suppose that a sequence $\{f_n\}$ consists of the functions:

$$f_n(x) = \begin{cases} 0 & \text{for } 0 \le x < \dfrac{1}{n+1}, \\ n^\alpha & \text{for } \dfrac{1}{n+1} \le x \le \dfrac{1}{n}, \\ 0 & \text{for } \dfrac{1}{n} < x \le 1. \end{cases} \tag{6.70}$$

The limit of the sequence reads as

$$f(x) \equiv \lim_{n\to\infty} f_n(x) = 0 \quad \text{almost everywhere,} \tag{6.71}$$

and thus

$$\int_0^1 \lim_{n\to\infty} f_n(x)dx = \int_0^1 f(x)dx = 0. \tag{6.72}$$

It also follows that

$$\int_0^1 f_n(x)dx = n^\alpha \int_{1/(n+1)}^{1/n} dx = n^\alpha \left(\frac{1}{n} - \frac{1}{n+1}\right) = \frac{n^{\alpha-1}}{n+1}, \tag{6.73}$$

which implies that

$$\lim_{n\to\infty} \int_0^1 f_n(x)dx = \begin{cases} 0 & (\alpha < 1), \\ 0 & (1 \le \alpha < 2), \\ 1 & (\alpha = 2), \\ \infty & (\alpha > 2). \end{cases} \tag{6.74}$$

But the story does not end. We should take care the case of $1 \leq \alpha < 2$; in this case, we cannot find any appropriate function g against f_n that satisfies $|f_n| \leq g$ for all n. Apparently, the following stepwise function g suffices because it dominates f_n for all n.

$$g(x) = \begin{cases} 1 & \text{for } \frac{1}{2} < x \leq 1, \\ 2^\alpha & \text{for } \frac{1}{3} < x \leq \frac{1}{2}, \\ 3^\alpha & \text{for } \frac{1}{4} < x \leq \frac{1}{3}, \\ \vdots & \vdots \\ n^\alpha & \text{for } \frac{1}{n+1} < x \leq \frac{1}{n}, \\ \vdots & \vdots \end{cases} \tag{6.75}$$

However, this function g is not integrable on $[0,1]$ if $1 \leq \alpha < 2$. In fact, we have

$$\int_0^1 g(x)dx = \sum_{n=1}^{\infty} \frac{n^{\alpha-1}}{n+1} = \begin{cases} 0 & (\alpha < 1), \\ \infty & (1 \leq \alpha < 2), \\ \infty & (\alpha \geq 2). \end{cases} \tag{6.76}$$

Therefore, the dominated convergence theorem fails for $1 \leq \alpha < 2$, although the *conclusion* of the theorem itself is true even when $1 \leq \alpha < 2$.

6.5 L^p space

6.5.1 *Essence of L^p space*

We close this chapter by demonstrating the direct relevance of the Lebesgue integral theory to functional analysis. The theory enables us to introduce a certain kind of functional spaces, called the L^p space. The space is spanned by complex-valued functions f such that $|f|^p$ is Lebesgue integrable. We already learned that L^2 spaces, which entered in section 3.3.5, belong to a specific class of L^p spaces with $p = 2$. Hence, every L^2 space is a member of **Hilbert spaces**. When $p \neq 2$, the L^p spaces lose the property of inner product, thus being excluded from Hilbert spaces. Nevertheless, the L^p spaces with $p \neq 2$ remain to be complete (i.e., they are **Banach spaces**). This is a main reason why the L^p spaces warrant attention in the study of functional analysis.

In the following discussion, we begin with the definition of L^p spaces from the viewpoint of measure, followed by examining how the spaces possess Banach space properties. Let p be a positive real number, and let X be a measurable set in $\mathbb{R}$. The L^p spaces is defined as follows:

Definition (L^p space):
The L^p space is such a set of complex-valued Lebesgue measurable functions $f(x)$ on X that satisfy for $p \geq 1$,

$$\int_X |f|^p d\mu < \infty. \quad \blacksquare \tag{6.77}$$

When the integral $\int_X |f(x)|^p dx$ exists, the p-root of the integral is called by the **p-norm** of f, denoted by

$$\|f\|_p = \left(\int_X |f|^p d\mu \right)^{1/p}. \tag{6.78}$$

Clearly for $p = 2$, the present definition reduces to our earlier definition of L^2.

6.5.2 Hölder's inequality

The following two inequalities are fundamentals that exhibit the relations between the norms of functions involved in L^p.

Theorem (Hölder's inequality):
Consider $f, g \in L^p$ that satisfy the conditions:

$$p, q > 1 \quad \text{and} \quad \frac{1}{p} + \frac{1}{q} = 1. \tag{6.79}$$

Then we have

$$fg \in L^1 \quad \text{and} \quad \|fg\|_1 \leq \|f\|_p \|g\|_q. \quad \blacksquare \tag{6.80}$$

Proof:
We assume that neither f nor g is zero **almost everywhere** (otherwise, the result is trivial). To proceed the proof, we first observe the inequality

$$a^{1/p} b^{1/q} \leq \frac{a}{p} + \frac{b}{q}, \quad \text{for } a, b \geq 0. \tag{6.81}$$

This inequality is justified by rewriting it as

$$t^{1/p} \leq \frac{t}{p} + \frac{1}{q}, \tag{6.82}$$

where we set $t = a/b$. Notice that the function given by

$$f(t) = t^{1/p} - \frac{t}{p} - \frac{1}{q}, \tag{6.83}$$

has a maximum at $t = 1$, namely,

$$\max f(t) = f(1) = 1 - \frac{1}{p} - \frac{1}{q} = 0. \tag{6.84}$$

We thus conclude that $f(t) \leq 0$ for every t, which results in the inequality (6.81). Next we employ the result to obtain

$$\frac{|f(x)g(x)|}{AB} \leq \frac{A^{-p}|f(x)|^p}{p} + \frac{B^{-q}|g(x)|^q}{q}, \tag{6.85}$$

where

$$A = \left[\int_X |f|^p d\mu\right]^{1/p} \quad \text{and} \quad B = \left[\int_X |g|^q d\mu\right]^{1/q}. \tag{6.86}$$

The right side of (6.85) is Lebesgue integrable due to the hypothesis that $f, g \in L^p$. Therefore, we obtain using (6.85)

$$\begin{aligned} \frac{1}{AB}\int_X |fg| d\mu &\leq \frac{A^{-p}}{p}\int_X |f|^p d\mu + \frac{B^{-q}}{q}\int_X |g|^q d\mu \\ &= \frac{1}{p} + \frac{1}{q} = 1. \end{aligned}$$

Consequently, we have

$$\int_X |fg| d\mu \leq AB, \tag{6.87}$$

which completes the proof. ■

6.5.3 *Minkowski's inequality*

The other inequality of interest is stated below.

Theorem (Minkowski's inequality):
If $f, g \in L^p$ with $p \geq 1$, then

$$f + g \in L^p \quad \text{and} \quad \|f+g\|_p \leq \|f\|_p + \|g\|_p. \quad ■ \tag{6.88}$$

Proof:
For $p = 1$, the inequality is readily obtained by the **triangle inequality**. For $p > 1$, it follows that

$$\int_X |f+g|^p d\mu \leq \int_X |f+g|^{p-1}|f| d\mu + \int_X |f+g|^{p-1}|g| d\mu. \tag{6.89}$$

Let $q > 0$ be such that

$$\frac{1}{p} + \frac{1}{q} = 1. \tag{6.90}$$

Applying **Hölder's inequality** to each of the last two integrals in (6.89) and noting that $(p-1)q = p$, we have

$$\begin{aligned}\int_X |f+g|^p d\mu &\leq M\left[\int_X |f+g|^{(p-1)q} d\mu\right]^{1/q} \\ &= M\left[\int_X |f+g|^p d\mu\right]^{1/q},\end{aligned} \tag{6.91}$$

where M denotes the right side of the inequality (6.88) that we would like to prove. Now divide the extreme ends of the relation (6.91) by

$$\left[\int_X |f+g|^p d\mu\right]^{1/q} \tag{6.92}$$

to obtain the desired result. ■

It should be noticed that both **Hölder's inequality** and **Minkowski's inequality** do not hold for $0 < p < 1$. This is why we restrict ourselves to $p \geq 1$.

6.5.4 Completeness of L^p space

By virtue of the two inequalities discussed above, we can show the completeness properties of L^p space.

Theorem (Completeness of L^p spaces):
The space L^p is complete: that is, for any $f_n \in L^p$ satisfying

$$\lim_{n,m\to\infty} \|f_n - f_m\|_p = 0, \tag{6.93}$$

there exists $f \in L^p$ such that

$$\lim_{n\to\infty} \|f_n - f\|_p = 0. \quad ■ \tag{6.94}$$

Proof:

Let $\{f_n\}$ be a **Cauchy sequence** in L^p. Then, there is a natural number n_1 such that for all $n > n_1$, and we have

$$\|f_n - f_{n_1}\| < \frac{1}{2}. \tag{6.95}$$

By induction, after finding $n_{k-1} > n_{k-2}$, we find $n_k > n_{k-1}$ such that for all $n > n_k$ we have

$$\|f_n - f_{n_k}\| < \frac{1}{2^k}. \tag{6.96}$$

Then $\{f_{n_k}\}$ is a subsequence of $\{f_n\}$ which satisfies

$$\|f_{n_{k+1}} - f_{n_k}\| < \frac{1}{2^k} \tag{6.97}$$

or

$$\|f_{n_1}\| + \sum_{k=1}^{\infty} \|f_{n_{k+1}} - f_{n_k}\| = A < \infty. \tag{6.98}$$

Let

$$g_k = |f_{n_1}| + |f_{n_2} - f_{n_1}| + \cdots + |f_{n_{k+1}} - f_{n_k}|, \quad k = 1, 2, \cdots. \tag{6.99}$$

Then, by **Minkowski's inequality**,

$$\begin{aligned}
\int_X g_k^p(x) d\mu &= \int_X (|f_{n_1}| + |f_{n_2} - f_{n_1}| + \cdots + |f_{n_{k+1}} - f_{n_k}|)^p d\mu \\
&\leq \left(\|f_{n_1}\|_p + \sum_{k=1}^{\infty} \|f_{n_{k+1}} - f_{n_k}\| \right)^p \\
&\leq A^p < \infty.
\end{aligned}$$

Let $g = \lim_{k\to\infty} g_k$. Then $g^p = \lim_{k\to\infty} g_k^p$. By the **monotone convergence theorem** given in §6.2.2, we have

$$\int_X g^p d\mu = \lim_{k\to\infty} \int_X g_k^p d\mu < \infty. \tag{6.100}$$

This shows that g is in L^p, and hence

$$\int_X \left(|f_{n_1}| + \sum_{k=1}^{\infty} |f_{n_{k+1}} - f_{n_k}| \right)^p dx < \infty. \tag{6.101}$$

This implies that

$$|f_{n_1}| + \sum_{k=1}^{\infty} |f_{n_{k+1}} - f_{n_k}| \tag{6.102}$$

converges almost everywhere to a function $f \in L^p$.

It remains to prove that $\|f_{n_k} - f\| \to 0$ as $k \to \infty$. We first notice that

$$f(x) - f_{n_j}(x) = \sum_{k=j}^{\infty} \left[f_{n_{k+1}}(x) - f_{n_k}(x) \right]. \tag{6.103}$$

It follows that

$$\|f - f_{n_j}\| \leq \sum_{k=j}^{\infty} \|f_{n_k+1} - f_{n_k}\|_p < \sum_{k=j}^{\infty} \frac{1}{2^k} = \frac{1}{2^{j-1}}. \tag{6.104}$$

Therefore, $\|f - f_{n_j}\|_p \to 0$ as $j \to \infty$. Now

$$\|f_n - f\|_p \leq \|f_n - f_{n_k}\|_p + \|f_{n_k} - f\|_p, \tag{6.105}$$

where $\|f_n - f_{n_k}\|_p \to 0$ as $n \to \infty$ and $k \to \infty$ and hence $\|f_n - f\|_p = 0$ as $n \to \infty$. This shows that the Cauchy sequence $\{f_n\}$ converges to f in L^p. ∎

Chapter 7
Wavelet

7.1 Continuous wavelet analysis

7.1.1 What is wavelet?

This chapter covers the minimum ground for understanding the **wavelet analysis**. The concept of **wavelet** originates from the study of signal analysis. In analyzing time-varying signals, **Fourier transformation** is known to be an efficient tool, because it enables to extract oscillation components of the signal in the time domain as well as to evaluate the power spectrum in the frequency domain. In certain cases, however, it is desirable to analyze the data from simultaneous viewpoints of both time and frequency domains. It is the wavelet analysis that accommodates this demand.

The crucial advantage of wavelet analyses is that they allow complicated information contained in a signal to be decomposed into elementary functions, called **wavelets**. Wavelets are localized waveforms whose characteristic time durations and characteristic frequencies are controllable over wide ranges of time- and frequency-scales. This controllability makes it possible to reconstruction and artificial modulation of the original signal with high precision and efficiency.

In the following discussions, we first demonstrate what constitutes a wavelet, and then discuss how it is used in the transformation of a signal. The primary question would be what a wavelet is. The answer is given by:

Definition (Wavelet):
A **wavelet** is a real-valued[1] function $\psi(t)$ having localized waveform, which satisfies the following criteria:

1. The integral of $\psi(t)$ is zero: $\int_{-\infty}^{\infty} \psi(t)dt = 0$.

2. The square of $\psi(t)$ integrates to unity: $\int_{-\infty}^{\infty} \psi(t)^2 dt = 1$.

3. The Fourier transform $\Psi(\omega)$ of $\psi(t)$ satisfies the **admissibility condition** expressed by

$$C_\Psi \equiv \int_0^\infty \frac{|\Psi(\omega)|^2}{\omega} d\omega < \infty. \tag{7.1}$$

Here, C_Ψ is called the **admissibility constant**, whose value depends on the explicit t-dependence of $\psi(t)$. ■

Observe that the condition 2 above says that $\psi(t)$ has to deviate from zero at finite intervals of t. On the other hand, the condition 1 tells us that any deviation above zero must be canceled out by deviation below zero. Hence, $\psi(t)$ must oscillate across the t-axis, resembling a wave. The followings are the most important two examples of wavelets.

Example 1:
The **Haar wavelet** (See Fig. 7.1(a)):

$$\psi(t) \equiv \begin{cases} -\dfrac{1}{\sqrt{2}}, & -1 < t \le 0, \\ \dfrac{1}{\sqrt{2}}, & 0 < t \le 1, \\ 0, & \text{otherwise.} \end{cases} \quad ■ \tag{7.2}$$

Example 2:
The **mexican hat wavelet** (See Fig. 7.1(b)):

$$\psi(t) \equiv \frac{2}{\sqrt{3\sigma}\pi^{1/4}}\left(1 - \frac{t^2}{\sigma^2}\right) e^{-t^2/(2\sigma^2)}. \quad ■ \tag{7.3}$$

To form the mexican hat wavelet (7.3), we start with the Gaussian function with mean zero and variance σ^2:

$$f(t) \equiv \frac{e^{-t^2/(2\sigma^2)}}{\sqrt{2\pi\sigma^2}}. \tag{7.4}$$

[1] We restrict our attention only to real-valued wavelets, while it is possible to define the complex-valued wavelets.

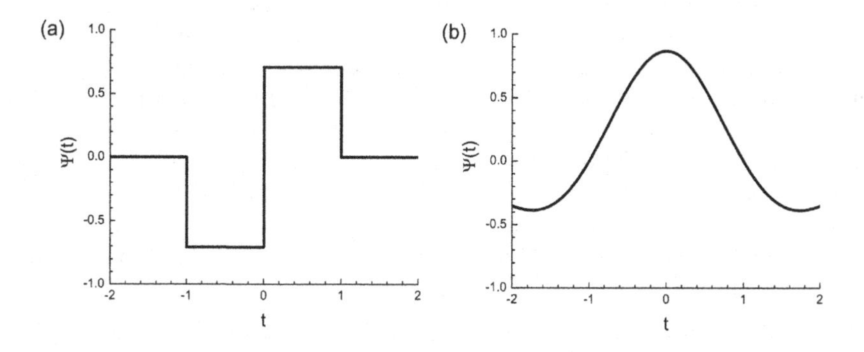

Figure 7.1: (a) The Haar wavelet given by (7.2). (b) The mexican hat wavelet given by (7.3) with $\sigma = 1$.

If we take the negative of the second derivative of $f(t)$ with normalization for satisfying the condition 2 noted above, we obtain the mexican hat wavelet.[2] In the meantime, we proceed our argument on the basis of the mexican hat wavelet with setting $\sigma = 1$ and omitting the normalization constant for simplicity.

Keypoint: A wavelet is a localized, oscillating wave; there are many varieties in the waveform.

7.1.2 Wavelet transform

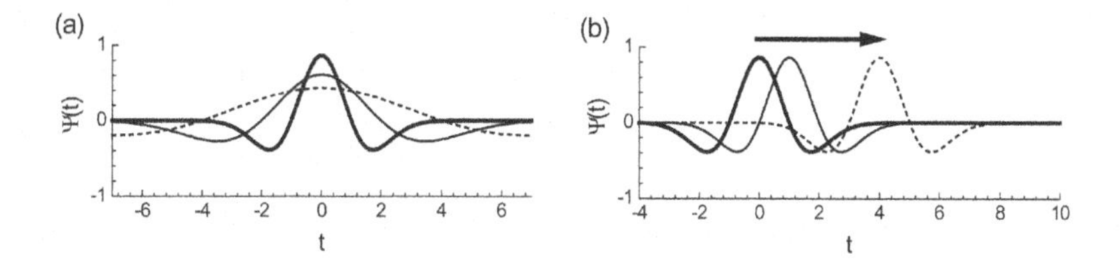

Figure 7.2: (a) Dilatation and (b) translation of a wavelet.

[2]More interestingly, all derivatives of the Gaussian function may be employed as a wavelet. Which is the most appropriate one to use depends on the application.

In mathematical terminology, the **wavelet transform** is a **convolution**[3] of a wavelet with a signal to be analyzed. The convolution involves two parameters that control the shape of the wavelet. The one parameter is the **dilatation parameter** denoted by a; it characterizes the dilation and contraction of the wavelet in the time domain (See Fig. 7.2 (a)). For the mexican hat wavelet, for instance, it is the distance between the center of the wavelet and its crossing of the time axis. The other parameter is the **translation parameter** b, which governs the movement of the wavelet along the time axis (Fig. 7.2 (b)). Using these notations, shifted and dilated versions of a mexican hat wavelet are expressed by

$$\psi\left(\frac{t-b}{a}\right) = \left[1-\left(\frac{t-b}{a}\right)^2\right]\exp\left[\frac{-1}{2}\cdot\left(\frac{t-b}{a}\right)^2\right], \tag{7.6}$$

where we have set $\sigma = 1$ in (7.3) and omitted the normalization factor for simplicity. We are now in a position to define the wavelet transform.

Definition (Wavelet transform):
The **wavelet transform** $T(a,b)$ of a continuous signal $x(t)$ with respect to the wavelet $\psi(t)$ is defined by

$$T(a,b) = w(a)\int_{-\infty}^{\infty} x(t)\psi\left(\frac{t-b}{a}\right)dt, \tag{7.7}$$

where $w(a)$ is an appropriate **weight function.** ■

Typically $w(a)$ is set to $1/\sqrt{a}$ because this choice yields

$$\int_{-\infty}^{\infty}\left[\frac{1}{\sqrt{a}}\psi\left(\frac{t-b}{a}\right)\right]^2 dt = \int_{-\infty}^{\infty}\psi(u)^2 du = 1 \quad \text{with } u = \frac{t-b}{a}, \tag{7.8}$$

i.e., the normalization condition as to the square integral of $\psi(t)$ remains invariant. This is why we use this value for the rest of this section.

The dilated and shifted wavelet is often written more compactly as

$$\psi_{a,b}(t) = \frac{1}{\sqrt{a}}\psi\left(\frac{t-b}{a}\right), \tag{7.9}$$

so that the transform integral may be written as

$$T(a,b) = \int_{-\infty}^{\infty} x(t)\psi_{a,b}(t)dt. \tag{7.10}$$

From now on, we will use this notation and refer to $\psi_{a,b}(t)$ simply as the wavelet.

[3]A **convolution** is an integral that expresses the amount of overlap of one function g as it is shifted over another function f. In usual, the convolution of two functions f and g over a finite range $[0,t]$ is written by

$$\int_0^{\tau} f(\tau)g(t-\tau)d\tau. \tag{7.5}$$

7.1.3 Correlation between wavelet and signal

Having defined the wavelet and its transform, we are ready to see how the transform is used as a signal analysis tool. In plain words, the wavelet transform works as a mathematical microscope, where b is the location on the time series being "viewed" and a represents the magnification at location b.

Keypoint: The wavelet transform works as a "microscope" for a time-varying signal.

Let us look at a simple example of evaluating the wavelet transform $T(a,b)$. Figures 7.3 and 7.4 show the same sinusoidal waves together with mexican hat wavelets of various locations and dilations. In Fig. 7.3 (a), the wavelet is located on a segment of the signal in which a positive part of the signal is fairly coincidental with that of the wavelet. This results in a large positive value of $T(a,b)$ given in (7.10). In Fig. 7.3 (b), the wavelet is moved to a new location where the wavelet and signal are out of phase. In this case, the convolution given by (7.10) produces a large negative value of $T(a,b)$. In between these two extrema, the value of $T(a,b)$ decreases from a maximum to a minimum as shown in Fig. 7.3 (c). The three figures thus clearly demonstrate how the wavelet transform $T(a,b)$ depends on the translation parameter b of the wavelet of interest.

In a similar way, Fig. 7.4 (a)-(c) show the dependence of $T(a,b)$ on the dilatation parameter a. When a is quite small, the positive and negative parts of the wavelet are all convolved by roughly the same part of the signal $x(t)$, producing a value of $T(a,b)$ near zero [See Fig. 7.4 (a)]. Likewise, $T(a,b)$ tends to zero as a becomes very large [See Fig. 7.4 (b)], since the wavelet covers many positive and negatively repeating parts of the signal. These latter two results indicate the fact that, when the dilatation parameter a is either very small or very large compared with the period of the signal, the wavelet transform $T(a,b)$ gives near-zero values.

7.1.4 Contour plot of wavelet transform

In practice, the wavelet transform plays an effective role in the signal analysis by systematically varying the values of the parameters a and b in $T(a,b)$ and displaying all the results in a contour plot. The left panel in Fig. 7.5 shows a contour plot of $T(a,b)$ versus a and b for a sinusoidal signal

$$x(t) = \sin t, \tag{7.11}$$

where the mexican hat wavelet has been used. The blight and shadowed regions indicate positive and negative magnitudes of $T(a,b)$, respectively. The near-zero values of $T(a,b)$ are evident in the plot at both large and small values of a. In addition, at intermediate values of a, we observe large undulations in $T(a,b)$ corresponding to the sinusoidal form of the signal. These wavelike behavior are

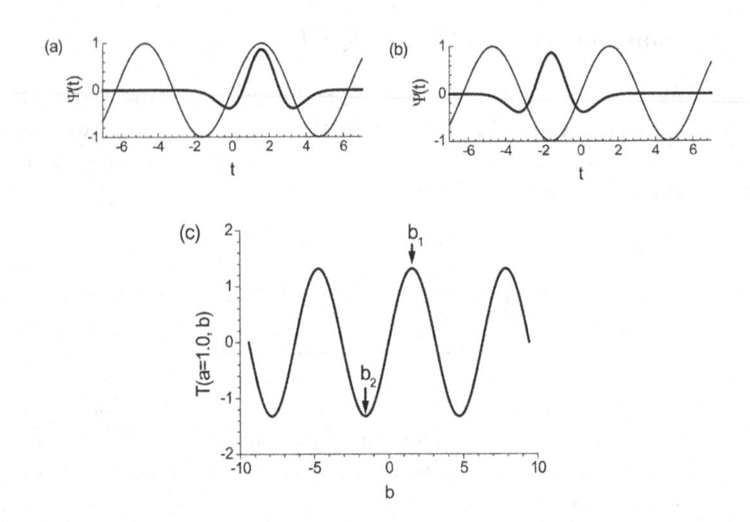

Figure 7.3: (a)-(b) Positional relations between the wavelet (thick) to signal (thin). The wavelet in (a) locating at $b_1 = \pi/2$ is in phase to signal, which results in a large positive value of $T(a,b)$ at b_1. In contrast, the wavelet in (b) locating at $b_2 = -\pi/2$ is out of phase to signal, which yields a large negative value of $T(b)$ at b_2. (c) The plot of $T(a = 1.0, b)$ as a function of b.

accounted for by referring back to Figs. 7.3 (a)-(b) and 7.4 (a)-(b), where wavelets move in and out of phase with the signal.

As summarized, when the wavelet matches the shape of the signal well at a specific scale and location, then a larger transform value is obtained. In contrast, if the wavelet and the signal do not correlate well, a low value of the transform is obtained. By performing such process at various locations of the signal and for various scales of the wavelet, we can investigate correlation between the wavelet and the signal.

Keypoint: The wavelet transform $T(a,b)$ of a signal $x(t)$ measure the degree of local correlation between the wavelet $\psi_{a,b}(t)$ and the signal.

The wavelet transformation procedure can be applied to signals that have more complicated wave form than a simple sinusoidal wave. The right panel in Fig. 7.5 shows a signal

$$x(t) = \sin t + \sin 3t \tag{7.12}$$

composed of two sinusoidal waves with different frequencies. The wavelet transform $T(a,b)$ of $x(t)$ is plotted in the right panel in Fig. 7.5. It is clear that the contribution from the wave with a higher frequency oscillation appears at a smaller a scale. This

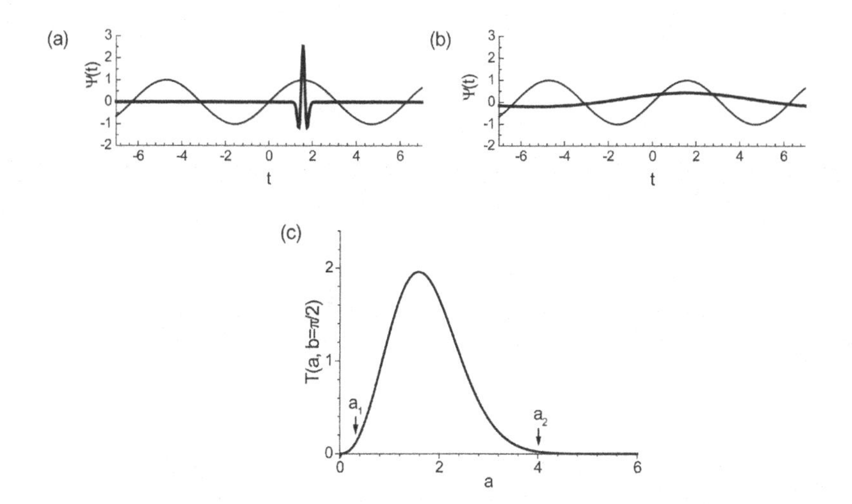

Figure 7.4: Wavelets with $a = 0.33$ (a) and $a = 4.0$ (b), in which $b = \pi/2$ is fixed. The resulting wavelet transform $T(a, b = \pi/2)$ as a function of a is given in (c).

clearly demonstrates the ability of the wavelet transform to decompose the original signal into its separate components.

7.1.5 Inverse wavelet transform

As similar to its Fourier counterpart, there is an **inverse wavelet transformation**, which enables us to reproduce the original signal $x(t)$ from its wavelet transform $T(a,b)$.

Theorem (Inverse wavelet transform):
If $x \in L^2(\mathbb{R})$, then f can be reconstructed by the formula:

$$x(t) = \frac{1}{C_\Psi} \int_{-\infty}^{\infty} db \int_0^{\infty} \frac{da}{a^2} T(a,b)\psi_{a,b}(t), \tag{7.13}$$

where the equality holds **almost everywhere**. ■

The proof of the formula (7.13) is based on the following theorem.

Theorem (Parseval's identity for wavelet transform):
Let $T_f(a,b)$ and $T_g(a,b)$ be the wavelet transform of $f, g \in L^2(\mathbb{R})$, respectively, associated with the wavelet $\psi_{a,b}(t)$. Then we have

$$\int_0^{\infty} \frac{da}{a^2} \int_{-\infty}^{\infty} db T_f(a,b) T_g^*(a,b) = C_\Psi \int_{-\infty}^{\infty} f(t) g(t)^* dt. \quad ■ \tag{7.14}$$

Proof (of the inverse wavelet transformation):

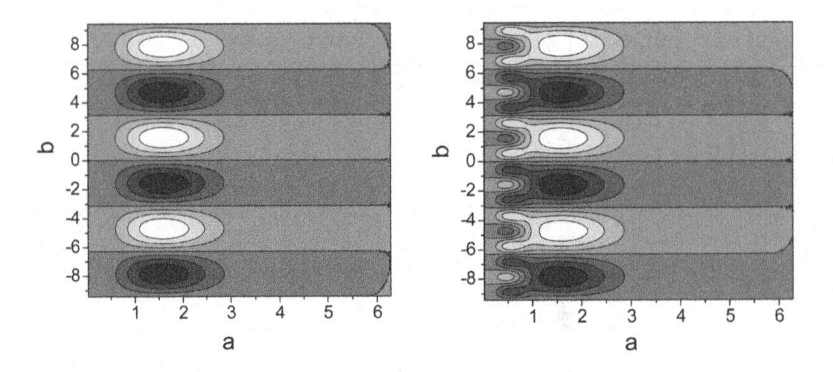

Figure 7.5: Left: Contour plot of the wavelet transform $T(a,b)$ of a sinusoidal wave $x(t) = \sin t$. Right: Contour plot of $T(a,b)$ of a complicated signal $x(t) = \sin t + \sin 3t$.

Suppose an arbitrary real function $g \in L^2(\mathbb{R})$. It follows from the Parseval identity that

$$\begin{aligned} C_\Psi \int_{-\infty}^{\infty} f(t)g(t)dt &= \int_{-\infty}^{\infty} db \int_0^{\infty} \frac{da}{a^2} T_f(a,b) T_g(a,b) \\ &= \int_{-\infty}^{\infty} db \int_0^{\infty} \frac{da}{a^2} T_f(a,b) \int_{-\infty}^{\infty} g(t)\psi_{a,b}(t)dt \\ &= \int_{-\infty}^{\infty} dt g(t) \left[\int_{-\infty}^{\infty} db \int_0^{\infty} \frac{da}{a^2} T_f(a,b)\psi_{a,b}(t) \right]. \end{aligned}$$

Since $g(t)$ is arbitrary, the inverse formula (7.13) follows. ■

7.1.6 Noise reduction

Using the inverse transform, we can reconstruct the original signal $x(t)$ from its transform $T(a,b)$. This is true no matter what degree of random noise perturbs the original signal. In view of practical use, however, it is convenient the noise contribution is removed from the original signal through the reconstruction procedure. Such **noise reduction** can be accomplished by a slight modulation of the inverse wavelet transform, as demonstrated below.

Suppose that we make the inverse transformation of $T(a,b)$ of the original signal $x(t)$ defined by

$$\tilde{x}(t) = \frac{1}{C_\Psi} \int_{-\infty}^{\infty} db \int_{a^*}^{\infty} \frac{da}{a^2} T(a,b)\psi_{a,b}(t). \tag{7.15}$$

The integration in the right side looks similar to that of the inverse transformation given by (7.13), except for the lower limit of the integration interval with respect to a. In the integration given in (7.15), the lower limit is set to be $a^* > 0$ instead of 0. This replacement in the lower limit causes $\tilde{x}(t)$ to deviate from the original signal $x(t)$, because of the loss of high-frequency components characterized by the parameter range with respect to a from $a = 0$ to $a = a^* > 0$. In practical use, this deviation property is made use of as a **noise reduction**.

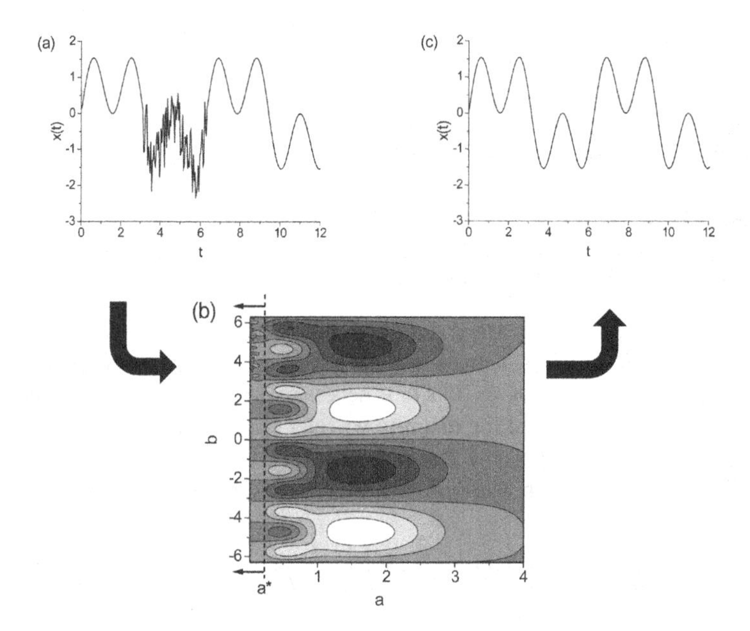

Figure 7.6: Noise reduction procedure through the wavelet transformation. (a) A signal $x(t) = \sin t + \sin 3t + R(t)$ with a local burst of noise $R(t)$. (b) The wavelet transform $T(a,b)$ of the $x(t)$. Noise reduction is performed through the inverse transformation of the $T(a,b)$ by applying an artificial condition of $T(a < a^*, b) = 0$. (c) The reconstructed signal $x^*(t)$ from the noise reduction procedure.

To demonstrate the noise reduction procedure, Fig. 7.6 (a) illustrates a segment of signal

$$x(t) = \sin t + \sin 3t + R(t) \tag{7.16}$$

constructed from two sinusoidal waveforms plus a local burst of noise $R(t)$. The wavelet transform of $x(t)$, denoted by $T(a,b)$, on the contour plot in Fig. 7.6 shows the two constituent waveforms at scales $a_1 = \pi/2$ and $a_2 = \pi/6$ in addition to a burst of noise around $b = 5.0$ at a high frequency region (i.e., small a scale).

Now we try to remove the high-frequency noise component by means of the noise reduction procedure. The elimination of noise is carried out by the inverse transform of $T(a,b)$ in which we artificially set $T(a,b) = 0$ for $a < a^*$. In effect we are reconstructing the signal using

$$x(t) = \frac{1}{C_\Psi} \int_{-\infty}^{\infty} db \int_{a^*}^{\infty} \frac{da}{a^2} T(a,b) \psi_{a,b}(t), \tag{7.17}$$

i.e., over a range of scales $[a^*, \infty)$. The lower integral limit, a^*, is the cut-off scale indicated by dotted line in Fig. 7.6 (b). As a result, the high frequency noise components evidently are reduced in the reconstructed signal as given in Fig. 7.6 (c).

Keypoint: Noise reduction technique enables to remove noise components that are localized at a specific time interval.

7.2 Discrete wavelet analysis

7.2.1 *Discrete wavelet transform*

In the previous section we learned the wavelet transform of signals that are continuous in time. In practice, signals are often represented by a series of decrete spikes, instead of continuous curves. It is thus necessary to develop a discrete version of the wavelet transforms, called the **discrete wavelet transform**.

We begin with the definition of a **discrete wavelet**. It is a natural generalization of a continuous wavelet, which was defined at scale a and location b by

$$\psi_{a,b}(t) = \frac{1}{\sqrt{a}}\psi\left(\frac{t-b}{a}\right), \tag{7.18}$$

in which the values of parameters a and b can change continuously. We now want to discretize values of a and b. A possible way is to use a logarithmic discretization of the a scale and link this to the size of steps taken between b locations. This kind of discretization yields

$$\psi_{m,n}(t) = \frac{1}{\sqrt{a_0^m}}\psi\left(\frac{t-nb_0a_0^m}{a_0^m}\right), \tag{7.19}$$

where the integers m and n control the wavelet dilation and translation respectively; a_0 is a specified fixed dilation step parameter, and b_0 is the location parameter. In the expression (7.19), the size of the translation steps, $\Delta b = b_0 a_0^m$, is directly proportional to the wavelet scale, a_0^m.

Common choices for discrete wavelet parameters a_0 and b_0 are $1/2$ and 1, respectively. This power-of-two logarithmic scaling of the dilation steps is known as the **dyadic grid arrangement**. Substituting $a_0 = 1/2$ and $b_0 = 1$ into (7.19), we obtain the **dyadic grid wavelet** represented by

$$\psi_{m,n}(t) = 2^{m/2}\psi(2^m t - n). \tag{7.20}$$

Using the dyadic grid wavelet of (7.20), we arrive at the **discrete wavelet transform** of a continuous signal $x(t)$ as follows.

Definition (Discrete wavelet transform):

$$T_{m,n} = \int_{-\infty}^{\infty} x(t)\psi_{m,n}(t)dt \text{ with } \psi_{m,n}(t) = \psi(2^m t - n)\,dt. \quad \blacksquare \tag{7.21}$$

Note that the discrete wavelet transform (7.21) differs from the discretized approximations of the continuous wavelet transform given by

$$T(a,b) = \int_{-\infty}^{\infty} x(t)\psi_{a,b}^{*}(t)dt \simeq \sum_{l=-\infty}^{\infty} x(l\Delta t)\psi_{a,b}^{*}(l\Delta t)\Delta t. \tag{7.22}$$

The latter involves arbitrary chosen values of a and b, while the wavelet transform $T_{m,n}$ is computed.

7.2.2 Complete orthonormal wavelet

The fundamental question is whether the original signal $x(t)$ can be constructed from the discrete wavelet transform $T_{m,n}$ through the relation as

$$x(t) = \sum_{m=-\infty}^{\infty}\sum_{n=-\infty}^{\infty} T_{m,n}\psi_{m,n}(t). \tag{7.23}$$

As intuitively understood, the reconstruction formula (7.23) is justified if the discretized wavelets $\psi_{m,n}(t)$ has the **orthonormality** and **completeness** property. On the one hand, the completeness of $\psi_{m,n}(t)$ implies that any function $x \in L^2(\mathbb{R})$ can be expanded by

$$x(t) = \sum_{m=-\infty}^{\infty}\sum_{n=-\infty}^{\infty} c_{m,n}\psi_{m,n}(t) \tag{7.24}$$

with appropriate expansion coefficients $c_{m,n}$. On the other hand, the orthonormality of $\psi_{m,n}(t)$ is represented by

$$\int_{-\infty}^{\infty} \psi_{m,n}(t)\psi_{m',n'}(t)dt = \delta_{m,n}\delta_{m',n'}. \tag{7.25}$$

Therefore, if $\psi_{m,n}(t)$ is endowed with the two properties, the coefficients $c_{m,n}$ in (7.24) read as

$$c_{m,n} = T_{m,n}, \tag{7.26}$$

because

$$\begin{aligned}
T_{m,n} &= \int_{-\infty}^{\infty} x(t)\psi_{m,n}(t)dt = \int_{-\infty}^{\infty}\left[\sum_{m'=-\infty}^{\infty}\sum_{n'=-\infty}^{\infty} c_{m',n'}\psi_{m',n'}(t)\right]\psi_{m,n}(t)dt \\
&= \sum_{m'=-\infty}^{\infty}\sum_{n'=-\infty}^{\infty} c_{m',n'}\int_{-\infty}^{\infty}\psi_{m,n}(t)\psi_{m',n'}(t)dt \\
&= \sum_{m'=-\infty}^{\infty}\sum_{n'=-\infty}^{\infty} c_{m',n'}\delta_{m,n}\delta_{m',n'} = c_{m,n}.
\end{aligned}$$

In general, however, the wavelets $\psi_{m,n}(t)$ given by (7.19) does neither have orthonormality nor completeness property. We thus achieve the following theorem.

Theorem (Validity of the inverse transformation formula):
The inverse transformation formula (7.23) is valid only for a limited class of sets of discrete wavelets $\{\psi_{m,n}(t)\}$ that is endowed with both orthonormality and completeness. ■

The simplest example of such desired wavelets is the **Haar discrete wavelet** presented below.

Example:
The Haar discrete wavelet is defined by

$$\psi_{m,n}(t) = 2^{m/2}\psi(2^m t - n), \tag{7.27}$$

where

$$\psi(t) = \begin{cases} 1 & 0 \le t < 1/2, \\ -1 & 1/2 \le t < 1, \\ 0 & \text{otherwise.} \end{cases} \tag{7.28}$$

This wavelet is known to be orthonormal and complete; its orthonormality is verified by the following discussion.

First we note that the norm of $\psi_{m,n}(x)$ is unity:

$$\begin{aligned} \int_{-\infty}^{\infty} \psi_{m,n}(t)^2 dt &= 2^{-m}\int_{-\infty}^{\infty} \left[\psi_{m,n}(2^{-m}t - n)\right]^2 dt \\ &= 2^{-m}\cdot 2^m \int_{-\infty}^{\infty} \psi_{m,n}(u)^2 du = 1. \end{aligned}$$

Thus, we obtain

$$\begin{aligned} I &\equiv \int_{-\infty}^{\infty} \psi_{m,n}(t)\psi_{k,l}(t)dt = \int_{-\infty}^{\infty} 2^{-m/2}\psi(2^{-m}t - n)2^{-k/2}\psi(2^{-k}t - \ell)dt \\ &= 2^{-m/2}.2^m \int_{-\infty}^{\infty} \psi(u)2^{-k/2}\psi[2^{m-k}(u+n) - \ell]dt \end{aligned} \tag{7.29}$$

If $m = k$, the integral in the last line in (7.29) reads

$$\int_{-\infty}^{\infty} \psi(u)\psi(u+n-\ell)dt = \delta_{0,n-\ell} = \delta_{n,\ell}, \tag{7.30}$$

since $\psi(u) \neq 0$ in $0 \le u \le 1$ and $\psi(u+n-\ell) \neq 0$ in $\ell - n \le u \le \ell - n + 1$ so that these intervals are disjointed from each other unless $n = \ell$. If $m \neq k$, it suffices to see the case of $m > k$ due to the symmetry. Put $r = m - k \neq 0$ in (7.29) to obtain

$$I = 2^{r/2}\int_{-\infty}^{\infty} \psi(u)\psi(2^r v + s)du = 2^{r/2}\left[\int_0^{1/2} \psi(2^r v + s)du - \int_{1/2}^{1} \psi(2^r v + s)du\right]. \tag{7.31}$$

This can be simplified as

$$I = \int_s^a \psi(x)dx - \int_a^b \psi(x)dx = 0, \tag{7.32}$$

where $2^r u + s = x$, $a = s + 2^{r-1}$ and $b = s + 2^r$. Observe that $[s,a]$ contains the interval $[0,1]$ of the Haar wavelet $\psi(t)$; this implies the first integral in (7.32) vanishes. Similarly, the second integral equals zero. It is thus concluded that

$$I = \int_{-\infty}^{\infty} \psi_{m,n}(t)\psi_{k,\ell} dt = \delta_{m,k}\delta_{n,\ell}, \tag{7.33}$$

which means that the Haar discrete wavelet $\psi_{m,n}(t)$ is orthonormal. ■

7.3 Wavelet space

7.3.1 *Multiresolution analysis*

We have known in section 7.2.2 that, for actual use of the inverse transform (7.23), we must find an appropriate set of discrete wavelets $\{\psi_{m,n}\}$ that possesses both orthonormality and completeness. In the remainder of this section, we describe the framework of constructing such discrete wavelets, which is based on the concept of **multiresolution analysis.**

The multiresolution analysis is a particular class of a set of function spaces.[4] It establishes a nesting structure of subspaces of $L^2(\mathbb{R})$, which allows us to construct a complete orthonormal set of functions (i.e., an **orthonormal basis**) for $L^2(\mathbb{R})$. The resulting orthonormal basis is nothing but the discrete wavelet $\psi_{m,n}(t)$ that satisfies the reconstruction formula (7.23). A rigorous definition of the multiresolution analysis is given below.

Definition (Multiresolution analysis):
A multiresolution analysis is a set of function spaces that consists of a sequence $\{\mathscr{V}_j : j \in \mathbb{Z}\}$ of closed subspaces of $L^2(\mathbb{R})$. Here the subspaces $\mathscr{V}_j$ satisfy the following conditions:

1. $\cdots \subset \mathscr{V}_{-2} \subset \mathscr{V}_{-1} \subset \mathscr{V}_0 \subset \mathscr{V}_1 \subset \mathscr{V}_2 \cdots \subset L^2(\mathbb{R})$,
2. $\bigcap_{j=-\infty}^{\infty} \mathscr{V}_j = \{0\}$,
3. $f(t) \in \mathscr{V}_j$ if and only if $f(2t) \in \mathscr{V}_{j+1}$ for all integer j,
4. There exists a function $\phi(t) \in \mathscr{V}_0$ such that the set $\{\phi(t-n),\ n \in \mathbb{Z}\}$ is an orthonormal basis for $\mathscr{V}_0$. ■

[4]Caution is required for the confusing terminology, that the multiresolution "analysis" is not an analytic method but "a set of function spaces".

The function $\phi(t)$ introduced above is called the **scaling function** (or **father wavelet**). It should be emphasized that the above definition gives no information as to the existence of (or the way to construct) the function $\phi(t)$ satisfying the condition **4**. However, once we find such desired function $\phi(t)$, we can establish the multiresolution analysis $\{\mathscr{V}_j\}$ by defining the function space $\mathscr{V}_0$ spanned by the orthonormal basis $\{\phi(t-n),\ n \in \mathbb{Z}\}$, and then forming other subspaces $\mathscr{V}_j$ $(j \neq 0)$ successively by using the property denoted in the condition 3 above.[5] If this is achieved, we say that *our scaling function $\phi(t)$ generates the multiresolution analysis* $\{\mathscr{V}_j\}$.

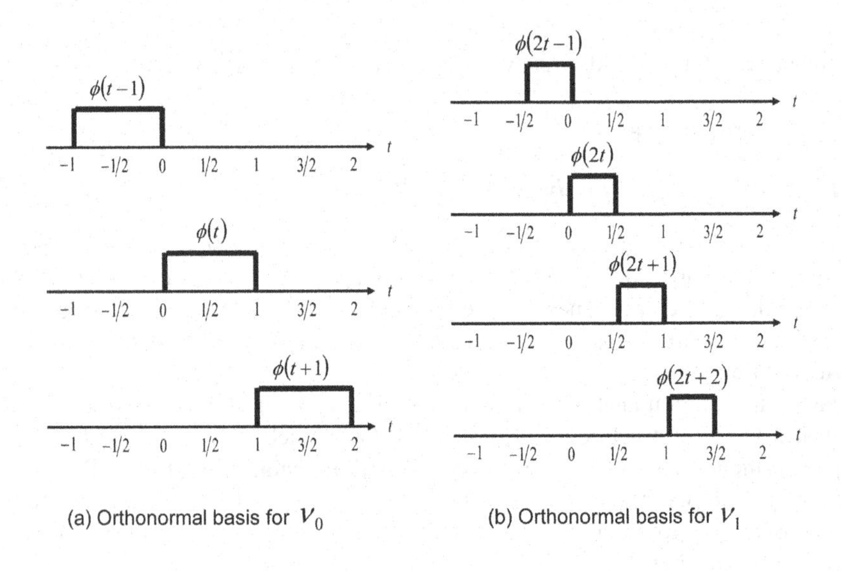

Figure 7.7: Two different sets of functions: $\mathscr{V}_0$ and $\mathscr{V}_1$.

> **Keypoint:** A multiresolution analysis is a sequence of "nested" subspaces of $L^2(\mathbb{R})$.

Example:
Consider the space $\mathscr{V}_m$ of all functions in $L^2(\mathbb{R})$ which are constant on each interval $[2^{-m}n, 2^{-m}(n+1)]$ for all $n \in \mathbb{Z}$. Obviously, the space $\mathscr{V}_m$ satisfies the conditions

[5] There is no straightforward way to construct a scaling function $\phi(t)$, or equivalently, a multiresolution analysis $\{\mathscr{V}_j\}$. Nevertheless, many kinds of scaling functions have been thus far discovered by means of sophisticated mathematical technique. We omit here the details of their derivations and just refer to the resulting scaling function at need.

from 1 to 3 of a multiresolution analysis. Furthermore, it is easy to see that the set $\{\phi(t-n),\ n \in \mathbb{Z}\}$ depicted in Fig. 7.7, which is defined by

$$\phi(t) = \begin{cases} 1, & 0 \le t \le 1 \\ 0, & \text{otherwise} \end{cases} \tag{7.34}$$

satisfies the condition **4**. Hence, any function $f \in \mathscr{V}_0$ can be expressed by

$$f(t) = \sum_{n=-\infty}^{\infty} c_n \phi(t-n), \tag{7.35}$$

with appropriate constants c_n. Thus, the spaces $\mathscr{V}_m$ consists of the multiresolution analysis generated by the scaling function (7.34). ■

7.3.2 Orthogonal decomposition

The importance of a multiresolution analysis lies in its ability to construct an orthonormal basis (i.e., a complete orthonormal set of functions) for $L^2(\mathbb{R})$. In order to prove this statement, we first remind that a multiresolution analysis $\{\mathscr{V}_j\}$ satisfies the relation:

$$\mathscr{V}_0 \subset \mathscr{V}_1 \subset \mathscr{V}_2 \subset \cdots \subset L^2. \tag{7.36}$$

We now define a space $\mathscr{W}_0$ as the **orthogonal complement** of $\mathscr{V}_0$ and $\mathscr{V}_1$, which yields

$$\mathscr{V}_1 = \mathscr{V}_0 \oplus \mathscr{W}_0. \tag{7.37}$$

Here, the symbol $\oplus$ indicates to take the **direct sum** of the given vector spaces. The space $\mathscr{W}_0$ we have introduced is called the **wavelet space** of the zero order; the reason of the name will be clarified in §7.3.3. The relation (7.37) extends to

$$\mathscr{V}_2 = \mathscr{V}_1 \oplus \mathscr{W}_1 \ = \mathscr{V}_0 \oplus \mathscr{W}_0 \oplus \mathscr{W}_1, \tag{7.38}$$

or, more generally, it gives

$$L^2 = \mathscr{V}_\infty \ = \mathscr{V}_0 \oplus \mathscr{W}_0 \oplus \mathscr{W}_1 \oplus \mathscr{W}_2 \oplus \cdots, \tag{7.39}$$

where $\mathscr{V}_0$ is the initial space spanned by the set of functions $\{\phi(t-n),\ n \in \mathbb{Z}\}$. Figure 7.8 pictorially shows the nesting structure of the spaces $\mathscr{V}_j$ and $\mathscr{W}_j$ for different scales j.

Since the scale of the initial space is arbitrary, it could be chosen at a higher resolution such as

$$L^2 = \mathscr{V}_5 \oplus \mathscr{W}_5 \oplus \mathscr{W}_6 \oplus \cdots, \tag{7.40}$$

or at a lower resolution such as

$$L^2 = \mathscr{V}_{-3} \oplus \mathscr{W}_{-3} \oplus \mathscr{W}_{-2} \oplus \cdots, \tag{7.41}$$

or at even negative infinity where (7.39) becomes

$$L^2 = \cdots \oplus \mathscr{W}_{-1} \oplus \mathscr{W}_0 \oplus \mathscr{W}_1 \oplus \cdots. \tag{7.42}$$

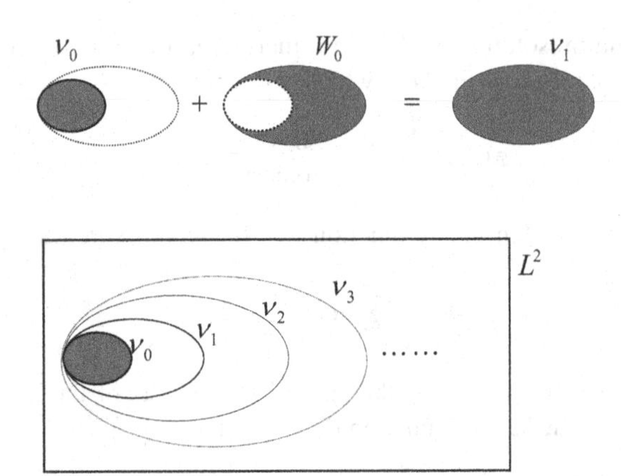

Figure 7.8: Hierarchical structure of the spaces $\mathscr{V}_j$ and $\mathscr{W}_j$ as subspaces of L^2.

The expression (7.42) is referred to as the **orthogonal decomposition** of the L^2 space. This represents that any function $x \in L^2(\mathbb{R})$ can be decomposed into the infinite sum of $g_j \in \mathscr{W}_j$ as

$$x(t) = \cdots + g_{-1}(t) + g_0(t) + g_1(t) + \cdots. \tag{7.43}$$

7.3.3 Orthonormal basis construction

We look into more about the orthogonal property of the wavelet spaces $\{\mathscr{W}_j\}$. From (7.37) and (7.38), we have

$$\mathscr{W}_0 \subset \mathscr{V}_1 \text{ and } \mathscr{W}_1 \subset \mathscr{V}_2. \tag{7.44}$$

In view of the definition of the multiresolution analysis $\{\mathscr{V}_j\}$, it follows that

$$f(t) \subset \mathscr{V}_1 \iff f(2t) \subset \mathscr{V}_2, \tag{7.45}$$

and therefore,

$$f(t) \in \mathscr{W}_0 \iff f(2t) \in \mathscr{W}_1. \tag{7.46}$$

Furtheremore, the condition 4 given in §7.3.1 results in that

$$f(t) \in \mathscr{W}_0 \iff f(t-n) \in \mathscr{W}_0 \text{ for any } n \in \mathbb{Z}. \tag{7.47}$$

The two results (7.46) and (7.47) are ingredients for constructing an orthonormal basis of $L^2(R)$ that we want, as demonstrated below.

We first assume that there exists a function $\psi(t)$ that leads to an orthonormal basis $\{\psi(t-n),\ n \in \mathbb{Z}\}$ for the space $\mathscr{W}_0$. Then, by using the notation:

$$\psi_{0,n}(t) \equiv \psi(t-n) \in \mathscr{W}_0, \tag{7.48}$$

it follows from (7.46) and (7.47) that its scaled version defined by

$$\psi_{1,n}(t) = \sqrt{2}\psi(2t-n) \tag{7.49}$$

serves as an orthonormal basis for $\mathscr{W}_1$. The term $\sqrt{2}$ was introduced to keep the normalization condition:

$$\int_{-\infty}^{\infty} \psi_{0,n}(t)^2 dt = \int_{-\infty}^{\infty} \psi_{1,n}(t)^2 dt = 1. \tag{7.50}$$

By repeating the same procedure, we find that the function

$$\psi_{m,n}(t) = 2^{m/2}\psi(2^m t - n) \tag{7.51}$$

constitutes an orthonormal basis for the space $\mathscr{W}_m$. Applying these results to the expression (7.43), we have for any $x \in L^2(\mathbb{R})$,

$$\begin{aligned} x(t) &= \cdots + g_{-1}(t) + g_0(t) + g_1(t) + \cdots \\ &= \cdots + \sum_{n=-\infty}^{\infty} c_{-1,n}\psi_{-1,n}(t) + \sum_{n=-\infty}^{\infty} c_{0,n}\psi_{0,n}(t) + \sum_{n=-\infty}^{\infty} c_{1,n}\psi_{1,n}(t) + \cdots \\ &= \sum_{m=-\infty}^{\infty}\sum_{n=-\infty}^{\infty} c_{m,n}\psi_{m,n}(t). \end{aligned} \tag{7.52}$$

Hence, the family $\psi_{m,n}(t)$ represents an orthonormal basis for $L^2(\mathbb{R})$. The above arguments are summarized by the following theorem.

Theorem:
Let $\{\mathscr{V}_j\}$ be a multiresolution analysis, and define the space $\mathscr{W}_0$ by $\mathscr{W}_0 = \mathscr{V}_1/\mathscr{V}_0$. If such a function $\psi(t)$ that leads to an orthonormal basis $\{\psi(t-n),\ n \in \mathbb{Z}\}$ for $\mathscr{W}_0$ is found, then the set of the functions $\{\psi_{m,n},\ m,n \in \mathbb{Z}\}$ given by

$$\psi_{m,n}(t) = 2^{m/2}\psi(2^m t - n) \tag{7.53}$$

constitutes an orthonormal basis for $L^2(\mathbb{R})$. ■

Emphasis is put on the fact that, since $\psi_{m,n}(t)$ is orthonormal basis for $L^2(\mathbb{R})$, the coefficients $c_{m,n}$ in (7.52) equal to the discrete wavelet transform $T_{m,n}$ given by (7.21). (See §7.2.2.) Therefore, the function $\psi(t)$ we introduce here is identified with the wavelet in the framework of continuous and discrete wavelet analysis, such as the Haar wavelet and the mexican hat wavelet. In this sense, each $\mathscr{W}_m$ is referred to as the **wavelet space**, and the function $\psi(t)$ is sometimes called the **mother wavelet**.

7.3.4 Two-scale relation

The preceding argument suggests that an orthonormal basis $\{\psi_{m,n}\}$ for $L^2(\mathbb{R})$ can be constructed by specifying the explicit function form of the mother wavelet $\psi(t)$. The remained task is, therefore, to develop a systematic way to determine the mother wavelet $\psi(t)$ that leads to an orthonormal basis $\{\psi(t-n)\ n \in \mathbb{Z}\}$ for the space $\mathscr{W}_0 = \mathscr{V}_1/\mathscr{V}_0$ contained in a given multiresolution analysis. We shall see that such $\psi(t)$ can be found by looking into the properties of the scaling function $\phi(t)$; we should remind that $\phi(t)$ yields an orthonormal basis $\{\phi(t-n)\ n \in \mathbb{Z}\}$ for the space $\mathscr{V}_0$. In this context, the space $\mathscr{V}_j$ is sometimes referred to as the **scaling function space**.

In this subsection, we mention an important feature of the scaling function $\phi(t)$, called the **two-scale relation**, that plays a key role in constructing the mother wavelet $\psi(t)$ of a given multiresolution analysis. We have already known that all functions in $\mathscr{V}_m$ are obtained from those in $\mathscr{V}_0$ through a scaling by 2^m. Applying this result to the scaling function denoted by

$$\phi_{0,n}(t) \equiv \phi(t-n) \in \mathscr{V}_0, \tag{7.54}$$

it follows that

$$\phi_{m,n}(t) = 2^{m/2}\phi(2^m t - n), \quad m \in \mathbb{Z} \tag{7.55}$$

is an orthonormal basis for $\mathscr{V}_m$. In particular, since $\phi \in \mathscr{V}_0 \subset \mathscr{V}_1$ and $\phi_{1,n}(t) = \sqrt{2}\phi(2t-n)$ is an orthonormal basis for $\mathscr{V}_1$, $\phi(t)$ can be expanded by $\phi_{1,n}(t)$. This is formally stated in the following theorem.

Definition (Two-scale relation):
If the scaling function $\phi(t)$ generates a multiresolution analysis $\{\mathscr{V}_j\}$, it satisfies the recurrence relation:

$$\phi(t) = \sum_{n=-\infty}^{\infty} p_n \phi_{1,n}(t) = \sqrt{2}\sum_{n=-\infty}^{\infty} p_n \phi(2t-n), \tag{7.56}$$

where

$$p_n = \int_{-\infty}^{\infty} \phi(t)\phi_{1,n}(t)dt. \tag{7.57}$$

This recurrence equation is called the **two-scale relation**[6] of $\phi(t)$, and the coefficients p_n are called the **scaling function coefficients**. ■

Example:
Consider again the space $\mathscr{V}_m$ of all functions in $L^2(\mathbb{R})$ which are constant on intervals $[2^{-m}n, 2^{-m}(n+1)]$ with $n \in \mathbb{Z}$. This multiresolution analysis is known to be generated by the scaling function $\phi(t)$ of (7.34). Substituting (7.34) into (7.57), we obtain

$$p_0 = p_1 = \frac{1}{\sqrt{2}} \text{ and } p_n = 0 \text{ for } n \neq 0, 1. \tag{7.58}$$

Thereby, the two-scale relation reads

$$\phi(t) = \phi(2t) + \phi(2t-1). \tag{7.59}$$

This means that the scaling function $\phi(t)$ in this case is a linear combination of its translates as depicted in Fig. 7.9. ■

---

[6]The two-scale relation is also referred to as the **multiresolution analysis equation**, the **refinement equation**, or the **dilation equation**, depending on the context.

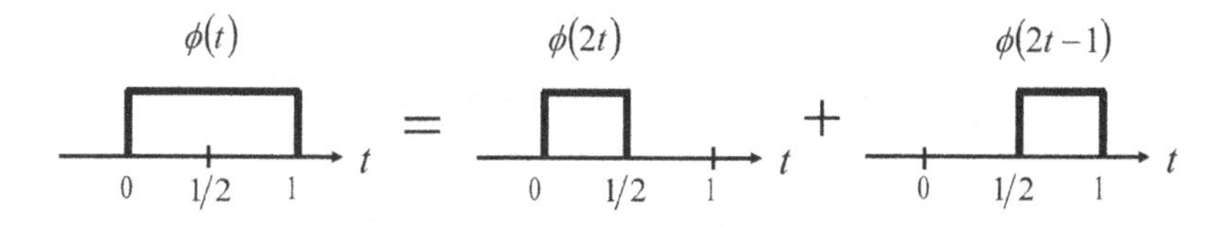

Figure 7.9: Two-scale relation of $\phi(t)$.

7.3.5 *Mother wavelet*

We are now in a position to determine the mother wavelet $\psi(t)$ that enables us to establish an orthonormal basis $\{\psi(t-n),\ n \in \mathbb{Z}\}$ for $L^2(\mathbb{R})$. Recall that a mother wavelet $\psi(t) = \psi_{0,0}(t)$ resides in a space $\mathscr{W}_0$ spanned by the next subspace of scaling function $\mathscr{V}_1$, that is, $\mathscr{W}_0 \subset \mathscr{V}_1$. Hence, in the same context as the previous subsection, $\psi(t)$ can be represented by a weighted sum of shifted scaling function $\phi(2t)$ by

$$\psi(t) = \sum_{n=-\infty}^{\infty} q_n \sqrt{2}\phi(2t-n), \quad n \in \mathbb{Z}. \tag{7.60}$$

The expansion coefficients q_n are called **wavelet coefficients**, given by

$$q_n = (-1)^{n-1} p_{-n-1} \tag{7.61}$$

as stated below.

Theorem:
If $\{\mathscr{V}_m\}$ is a multiresolution analysis with the scaling function $\phi(t)$, the mother wavelet $\psi(t)$ is given by

$$\psi(t) = \sqrt{2} \sum_{n=-\infty}^{\infty} (-1)^{n-1} p_{-n-1}\phi(2t-n), \quad n \in \mathbb{Z}, \tag{7.62}$$

where p_n is the scaling function coefficient of $\phi(t)$. ■

It is to be noted that p_n in (7.62) is uniquely determined by the function form of the scaling function $\phi(t)$; See (7.57). The above theorem states, therefore, that the mother wavelet $\psi(t)$ is obtained once specifying the scaling function $\phi(t)$ of a given multiresolution analysis.

7.3.6 *Multiresolution representation*

Through the discussions thus far, we have obtained an orthonormal basis consisting of scaling functions $\phi_{j,k}(t)$ and wavelets $\psi_{j,k}(t)$ that span all of $L^2(\mathbb{R})$. Since

$$L^2 = \mathscr{V}_{j_0} \oplus \mathscr{W}_{j_0} \oplus \mathscr{W}_{j_0+1} \oplus \cdots, \tag{7.63}$$

any function $x(t) \in L^2(\mathbb{R})$ can be expanded such as

$$x(t) = \sum_{k=-\infty}^{\infty} S_{j_0,k}\phi_{j_0,k}(t) + \sum_{k=-\infty}^{\infty}\sum_{j=j_0}^{\infty} T_{j,k}\psi_{j,k}(t). \tag{7.64}$$

Here, the initial scale j_0 could be zero or other integer, or negative infinity as in (7.23) where no scaling functions are used. The coefficients $T_{j,k}$ are identified with the discrete wavelet transform already given in (7.21). Often $T_{j,k}$ in (7.64) is called the **wavelet coefficient**, and correspondingly $S_{j,k}$ is called the **approximation coefficient**.

The representation (7.64) can be more simplified by using the following notations. We denote the first summation in the right of (7.64) by

$$x_{j_0}(t) = \sum_{k=-\infty}^{\infty} S_{j_0,k}\phi_{j_0,k}(t); \tag{7.65}$$

this is called the **continuous approximation** of the signal $x(t)$ at scale j_0. Observe that the continuous approximation approaches $x(t)$ in the limit of $j_0 \to \infty$, since in this case $L^2 = \mathscr{V}_\infty$. In addition, we introduce the notation that

$$z_j(t) = \sum_{k=-\infty}^{\infty} T_{j,k}\psi_{j,k}(t), \tag{7.66}$$

where $z_j(t)$ is called the **signal detail** at scale j. With these conventions, we can write (7.64) as

$$x(t) = x_{j_0}(t) + \sum_{j=j_0}^{\infty} z_j(t). \tag{7.67}$$

The expression (7.67) says that the original continuous signal $x(t)$ is expressed as a combination of its continuous approximation x_{j_0} at arbitrary scale index j_0, added to a succession of signal details $z_j(t)$ from scales j_0 up to infinity.

Additional noteworthy is that, due to the nested relation of $\mathscr{V}_{j+1} = \mathscr{V}_j \oplus W_j$, we can write

$$x_{j+1}(t) = x_j(t) + z_j(t). \tag{7.68}$$

This indicates that if we add the signal detail at an arbitrary scale (index j) to the continuous approximation at the same scale, we get the signal approximation at an increased resolution (i.e., at a smaller scale, index $j+1$). The important relation (7.68) between continuous approximations $x_j(t)$ and signal details $z_j(t)$ are called a **multiresolution representation**.

Chapter 8

Distribution

8.1 Motivation and merits

8.1.1 Overcoming the confusing concept: "sham function"

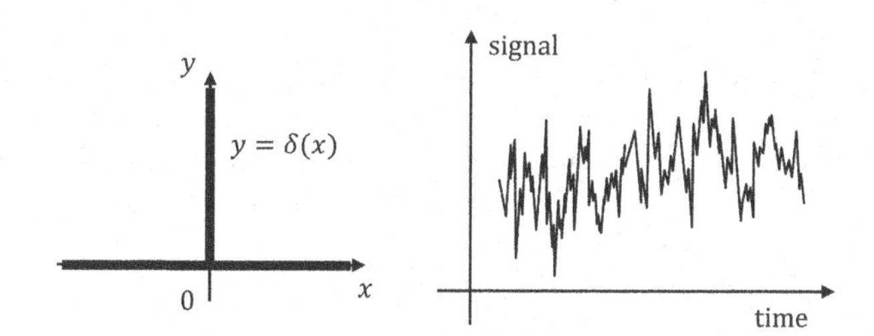

Figure 8.1: Examples of sham functions.

In different fields of physics and engineering, we sometimes deal with **sham functions**[1] that do not satisfy the axiom of functions in a rigorous sense. A typical example of a sham function is **Dirac's delta function**[2] $\delta(\boldsymbol{r})$. It is made up of, for instance, describing an electric point charge.[3] In physics and engineering, the spatial

[1] By the phrase "sham function," we mean an ill-defined mathematical object, which cannot be an exact function but its imitation.

[2] Dirac's delta function is commonly called a "function," even though it is not mathematically considered a genuine function.

[3] We should remind here that a point charge must be a point with no spatial extent, at which the electric charge should be concentrated. This situation cannot be described by a function in the usual sense.

distribution of a point charge $\rho(\boldsymbol{r})$ is usually expressed by

$$\rho(\boldsymbol{r}) = \rho_0 \delta(\boldsymbol{r}), \tag{8.1}$$

and $\delta(\boldsymbol{r})$ is defined by

$$\delta(\boldsymbol{r}) = 0 \text{ at } \boldsymbol{r} \neq \boldsymbol{0}, \qquad \iiint \delta(\boldsymbol{r}) d\boldsymbol{r} = 1. \tag{8.2}$$

However, the expression of $\delta(\boldsymbol{r})$ is not consistent with our familiar concept of functions (see section 8.2.1). In fact, the discrete and divergence properties inherent in $\delta(\boldsymbol{r})$ make it impossible to apply differentiation, integration, or Fourier transformation to $\delta(\boldsymbol{r})$. Hence, the sham function $\delta(\boldsymbol{r})$ should be of no use in describing the physical phenomena triggered by the presence or temporal movement of the point charge.[4] If this is the case, then why is $\delta(\boldsymbol{r})$ repeatedly used in physics and engineering textbooks?

Another example of a sham function is one used to describe a sequential line graph that has an infinite number of discrete jumps. This type of line graph is observed in financial data analysis and signal processing. Inconveniently, this type of extremely jagged data prevents us from using common elementary calculus tools, because data graphs in elementary calculus are assumed to be smooth enough or have a finite number of jumps at most. These difficulties can be avoided, in principle, if we restrict our consideration to continuous and sufficiently smooth functions. However, this intentional restriction destroys the versatility of the mathematical theory we are attempting to develop.

What helps to resolve the problems mentioned above is the concept of **distribution**.[5] In order to introduce the notion of distribution, we first need to expand the meaning of function (= mapping) from the classical, conventional one to more generalized counterpart. In the classical sense, functions play a role of:

"The mapping from a number to a number."

We now introduce a generalized alternative that plays a role of:

"The mapping from a (classical) function to a number."

Out task is not over yet. The next task is to pick up a limited kind of mappings among the latter generalized ones. The mappings to be picked up must satisfy certain conditions, as we will learn step-by-step in this chapter. The resulting specially chosen mappings that relate a classical function to a number are what we call by **distributions**.

> **Keypoint:** A distribution is a limited class of mappings that transform a function to a number.

[4] The phenomena include spatial profile of an electric potential field and electromagnetic induction.

[5] It is also called the **Schwartz distribution** after discoverer Laurent Schwartz.

8.1.2 Merits of introducing "Distribution"

Distributions are efficient tools that remove mathematical ambiguities, particularly in manipulating sham functions like Dirac's delta functions.

For instance, distributions make it possible to differentiate many functions whose derivatives do not exist in the classical sense. As well, distributions are widely used in the theory of partial differential equations, in which checking the existence of a solution is much more difficult when using classical methods.

Distributions are also important in physics and engineering. In these fields, many problems naturally lead to differential equations whose solutions or initial conditions involve discontinuity or divergence. Particular emphasis should be placed on the fact that using distributions to apply differentiation and Fourier transformation plays a powerful role in the study of ill-defined functions that are endowed with discontinuous jumps or divergent points.

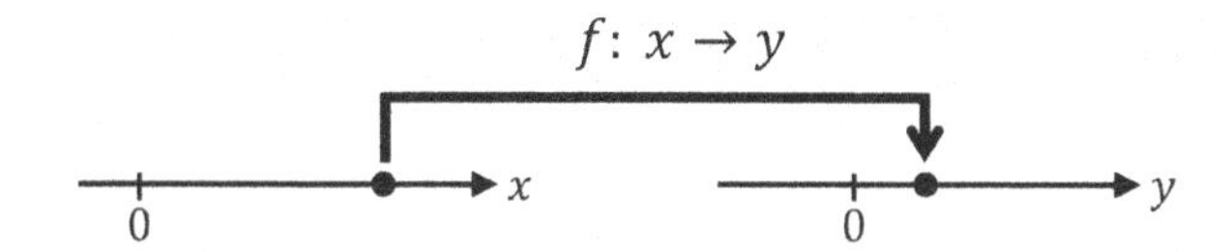

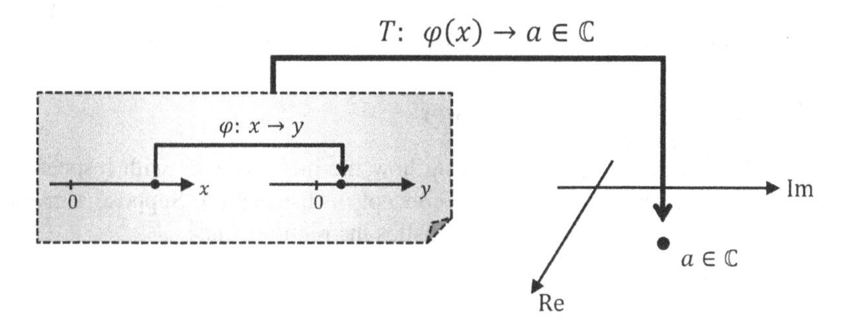

Figure 8.2: A function $f : x \to y$ and a distribution $T : \varphi(x) \to a \in \mathbb{C}$.

8.2 Establishing the concept of distribution

8.2.1 Inconsistency hidden in the δ-function

This section demonstrates the way of thinking how we can construct the concept of distribution. As the first step, let us compare classical (i.e., usual) functions with sham functions. Suppose that a real (or complex) number y is assigned to every real number x and these real numbers are densely distributed along the one-dimensional line $\mathbb{R}$. The rule that relates the two numbers is a function written as

$$f : x \to y, \tag{8.3}$$

and the notation $y = f(x)$ symbolizes the relation between x and y. In contrast to this normal function, certain sham functions, which are incompatible with our familiar concept of function, have been introduced into physics and engineering. As already mentioned, Dirac's delta function is the most common sham function and is defined by

$$\delta(x) = 0 \text{ for } x \neq 0, \quad \int_{-\infty}^{\infty} \delta(x)dx = 1. \tag{8.4}$$

The reason this is incompatible with our familiar concept of functions is because the measure[6] of an isolated point $x = 0$ is zero, i.e., $\int_0^0 f(x)dx = 0$ for arbitrary functions $f(x)$. This means that

$$\begin{aligned} \int_{-\infty}^{\infty} \delta(x)dx &= \int_{-\infty}^{0} \delta(x)dx + \int_{0}^{0} \delta(x)dx + \int_{0}^{\infty} \delta(x)dx \\ &= 0+0+0 = 0, \end{aligned} \tag{8.5}$$

which obviously contradicts the original definition of $\delta(x)$. Despite this self-defeating contradiction, $\delta(x)$ appears so frequently in applied sciences and is heavily used for practical calculations. The remarkable efficiency of such an ill-defined mathematical entity is due to the fact that $\delta(x)$ becomes a self-consistent, well-defined entity if we consider it as a distribution (not a classical function).

8.2.2 *How to resolve the inconsistency*

Demonstrated below is the way of thinking how the inconsistency with respect to $\delta(x)$ may be resolved by introducing the concept of distribution. Suppose a given mathematical entity, designated by $\delta(x)$, satisfies the requirements:

$$\delta(x) = 0 \ (x \neq 0), \quad \int_{-\infty}^{\infty} \delta(x)dx = 1. \tag{8.6}$$

If this were justified, the following equation should hold for an arbitrary function $\varphi(x)$,

$$\int_{-\infty}^{\infty} \delta(x)\varphi(x)dx = \varphi(0)\int_{-\infty}^{\infty} \delta(x)dx = \varphi(0). \tag{8.7}$$

In the derivation of the middle expression from the left one, we assume that[7] $\delta(x) = 0 \ (x \neq 0)$. We further assume that $\delta(x)$ is differentiable everywhere, including

---

[6]In mathematical terminology, a **measure** is a generalization of the concept of size, such as length, area, and volume. See section 6.2.1 in details.

[7]We factored out the constant $\varphi(0)$ from the integrand. We can do this because $\delta(x) = 0$ except at $x = 0$ and the integrand vanishes except at $x = 0$. Hence, we can pull a portion of the integrand at $x = 0$, $\varphi(0)$, outside of the integral.

the point $x = 0$. We designate the derivative of $\delta(x)$ by $\delta'(x)$ and perform partial integration to obtain

$$\int_{-\infty}^{\infty} \delta'(x)\varphi(x)dx = [\delta(x)\varphi(x)]_{x=-\infty}^{x=\infty} - \int_{-\infty}^{\infty} \delta(x)\varphi'(x)dx = -\varphi'(0). \tag{8.8}$$

Needless to say, neither $\delta(x)$ nor $\delta'(x)$ can be considered functions in a conventional sense. Therefore, these two equations do not make sense as they are.

In order to resolve this inconsistency, we need to invert our thinking. Let us regard δ not as a function that relates two numbers but as a mapping that satisfies (8.7). Namely, we define δ by the mapping that maps a given function $\varphi(x)$ to a specific number[8] $\varphi(0)$, and we write

$$\delta : \varphi(x) \to \varphi(0). \tag{8.9}$$

This expression clarifies the workings of δ. It acts as a pair of tweezers, which picks up $\varphi(0)$, i.e., the value of a given function $\varphi(x)$ at $x = 0$. The same understanding applies to δ', the derivative of δ. On the basis of (8.8), we regard it as the mapping represented by

$$\delta' : \varphi(x) \to -\varphi'(0), \tag{8.10}$$

which maps a given $\varphi(x)$ to the number $-\varphi'(0)$. This is generally what is meant when we say that Dirac's delta function is not a function but a distribution (or mapping).

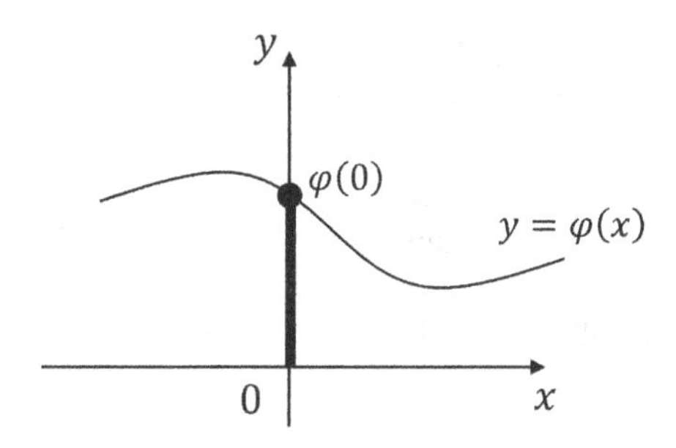

Figure 8.3: Workings of Dirac's delta function δ as a distribution.

Keypoint: In the sense of distribution, Dirac's delta function serves as a "tweezer" that picks up the number $\varphi(0)$ from the given function $\phi(x)$.

[8]This number is generally complex valued.

8.2.3 Definition of distribution

Now we are ready to enjoy the precise definition of distributions in general terms, those not limited to Dirac's delta function.

Definition (Distribution):
A **distribution** is a mapping that relates a function[9] $\varphi(x)$ to a complex number $a \in \mathbb{C}$. ■

We express this mapping as T and write the relation between $\varphi(x)$ and a as

$$T : \varphi(x) \to a \in \mathbb{C}. \tag{8.11}$$

It is customary to describe this relation using the symbol "$\langle\, ,\, \rangle$":

$$\langle T, \varphi \rangle = a \in \mathbb{C}. \tag{8.12}$$

8.3 Examples of distribution

8.3.1 Dirac's delta function

This section presents a few simple examples of distributions. The first to be noted is **Dirac's delta function**, defined by

$$\langle \delta, \varphi \rangle = \varphi(0) \in \mathbb{C} \tag{8.13}$$

Since δ is not a function in the usual sense, it is meaningless to say: "the value of $\delta(x)$ at x." However, things change drastically if we consider it to be a distribution. As a distribution, δ is a proper mathematical entity that transforms a function $\varphi(x)$ into the number $\varphi(0)$.

A slight generalization of δ is the distribution δ_c, defined by[10]

$$\langle \delta_c, \varphi \rangle = \varphi(c) \in \mathbb{C}. \tag{8.14}$$

Here, the subscript c in δ_c indicates that the value of $\varphi(x)$ is extracted at $x = c$ rather than at $x = 0$.

[9]We will see later that the function $\varphi(x)$, characterizing the nature of a distribution, must be a **rapidly decreasing function** with respect to x. Refer to section 8.4.1 for details.

[10]δ_c is also called Dirac's delta function because it is similar to δ.

8.3.2 Heaviside's step function

The next example is **Heaviside's step function**, commonly denoted by H. It works as the mapping defined by

$$\langle H, \varphi \rangle = \int_0^\infty \varphi(x)dx \in \mathbb{C}. \tag{8.15}$$

The distribution H transforms a function $\varphi(x)$ to the integral value of $\int_0^\infty \varphi(x)dx$. Note that this quantity is a number determined independently of x, although the integrand involves an x-dependent entity $\varphi(x)$.

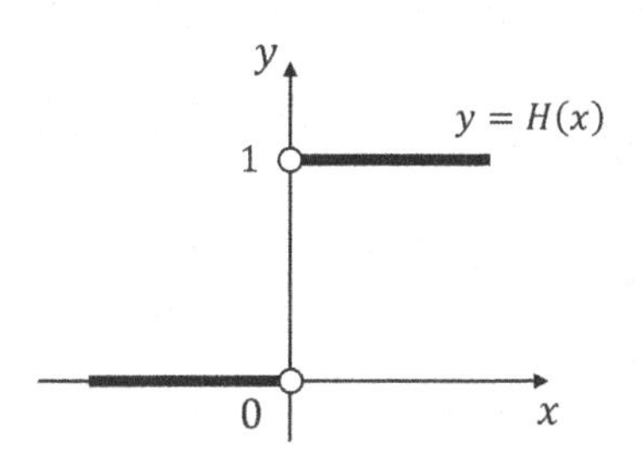

Figure 8.4: Heaviside's step function.

Interestingly, the distribution H is a counterpart to our familiar sham function $H(x)$, expressed by

$$H(x) = \begin{cases} 1, & x > 0 \\ 0, & x < 0 \end{cases}, \tag{8.16}$$

which frequently appears in physics and engineering. This sham function exhibits a discontinuity at $x = 0$, making it indifferentiable at this point in the usual sense. However, if we consider it as a distribution, then it becomes differentiable even at $x = 0$, as will be demonstrated (see section 8.5.1).

8.3.3 Cauchy's principal value as a distribution

Another important example of a distribution is $\text{PV}\frac{1}{x}$, called **Cauchy's principal value**, which is defined by

$$\left\langle \text{PV}\frac{1}{x}, \varphi \right\rangle = \lim_{\varepsilon \to +0} \int_{|x|>\varepsilon} \frac{\varphi(x)}{x} dx \in \mathbb{C}. \tag{8.17}$$

The notation used in (8.17) is rather complicated and requires a few supplemental remarks. First, PV is an acronym for principal value. Second, this symbol also appears in elementary analysis, in which it is used in the integral combination

$$\text{PV} \int_a^b f(x)dx \equiv \lim_{\varepsilon \to +0} \left[\int_a^{c-\varepsilon} f(x)dx + \int_{c+\varepsilon}^b f(x)dx \right]. \tag{8.18}$$

Note that the integral $\int_a^b f(x)dx$ does not always exist. In particular, when $f(x)$ is divergent at a point in the interval $[a,b]$, the integral $\int_a^b f(x)dx$ may diverge. Even in these cases, the right-hand side in (8.18) is sometimes convergent[11], depending on the behavior of $f(x)$. Under this circumstance, we say:

"The integral $\int_a^b f(x)dx$ is convergent
in the *classical* sense of Cauchy's principal value."

Here, PV $\int_a^b f(x)dx$ is its convergence value, which can be evaluated using the limit procedure given in (8.18). This is the concept of Cauchy's principal value as it applies to elementary analysis.

In contrast to the classical counterpart, Cauchy's principal value in the distributional sense is a mapping written by

$$\mathrm{PV}\frac{1}{x} : \varphi(x) \to \lim_{\varepsilon\to+0}\int_{|x|>\varepsilon} \frac{\varphi(x)}{x}dx. \tag{8.19}$$

Remember that the symbol x contained in the notation $\mathrm{PV}\frac{1}{x}$ is no longer a variable into which a certain number is substituted; it is an accessary with no meaning, and $\mathrm{PV}\frac{1}{x}$ as a whole indicates the mapping from $\varphi(x)$ to the number specified in (8.19).

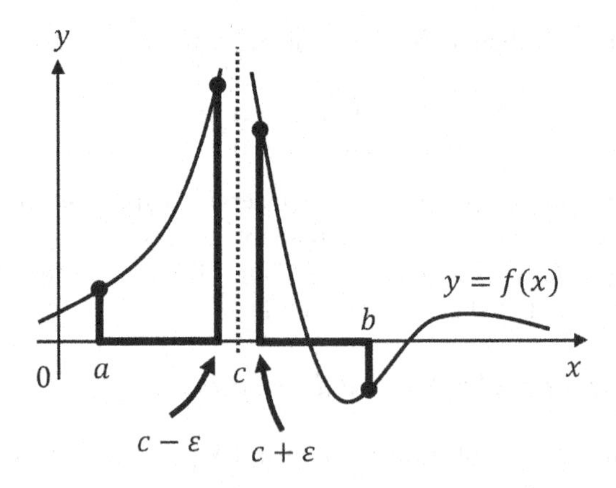

Figure 8.5: Cauchy's principal value.

[11] For instance, $\int_{-1}^{1}\frac{dx}{x}$ does not exist because the integrand $1/x$ diverges at $x=0$. Nevertheless, $\lim_{\varepsilon\to+0}\left[\int_{-1}^{-\varepsilon}\frac{dx}{x}+\int_{\varepsilon}^{1}\frac{dx}{x}\right]$ converges due to the symmetry of $1/x$ with respect to the point $x=0$. This is the reason we can define Cauchy's principal value PV $\int_{-1}^{1}\frac{dx}{x}$ in the classical sense.

8.3.4 Remarks on the distribution PV$\frac{1}{x}$

The reason why we introduce this distribution PV$\frac{1}{x}$ is to address the computational inconvenience inherent in the function $1/x$. The function $1/x$ is not differentiable at $x = 0$ and not integrable on $(-\infty\ ,\ +\infty)$. As a result, it is impossible to perform a Fourier transformation on $1/x$. The "exclusivity" of the function $1/x$ creates difficulties when solving physics or engineering problems, as was the case in our example of the electric potential field generated by a point charge. These difficulties can be resolved by applying the concept of a distribution. Here, the function $1/x$ is replaced by the distribution PV$\frac{1}{x}$, followed by differentiation, integration, and Fourier transformation as it pertains to the distribution. Using this procedure, we obtain the well-known results without mathematical ambiguity, as we will see later.

Keypoint: Use of the distribution PV$\frac{1}{x}$ enables us to make Fourier transform of the function $\frac{1}{x}$, though it is forbidden in the classical sense.

We should also note that when considering Cauchy's principal value in the distributional sense, the integration range should be from $-\infty$ to $+\infty$, except for $-\varepsilon < x < \varepsilon$, as expressed by

$$\int_{|x|>\varepsilon} \left[= \int_{-\infty}^{-\varepsilon} + \int_{\varepsilon}^{+\infty} \right]. \tag{8.20}$$

On the other hand, the classical understanding of Cauchy's principal value is free from this restriction. In the classical counterpart defined by (8.18), the integration range $\int_a^b dx$ is chosen arbitrarily[12] and is not necessarily $(-\infty,\ +\infty)$. This is an auxiliary difference between the classical and distributional approaches to Cauchy's principal value.

8.3.5 Properties of distribution

The major advantages of distributions are listed below.

1. Distributions are infinitely differentiable.
2. Distributions are differentiable at points where their classical counterparts are discontinuous or divergent.
3. Distributions allow Fourier transformations even when these do not exist for their classical counterparts.
4. Distributions provide a new class of solutions (physically reasonable) for a certain class of differential equations.

[12]For instance, we can write PV $\int_a^b \frac{dx}{x} = \log\left|\frac{b}{a}\right|$ for arbitrary $a(<0)$ and $b(>0)$. Here, the constants a and b do not need to be $\pm\infty$, which differs from the distribution case.

The properties 1 and 2 listed above are realized by our earlier discussion of δ', the derivative of **Dirac's delta function**. It is obvious that the sham function $\delta(x)$ cannot be differentiated at $x = 0$ in the classical sense. But surprisingly, δ as a distribution is infinitely differentiable even at $x = 0$, contrary to our intuitive understanding of its classical counterpart. Similarly, **Heaviside's step function** H as a distribution is infinitely differentiable everywhere, even though its classical counterpart (i.e., the sham function $H(x)$) shows discontinuous jump at $x = 0$. Furthermore, the derivative of H at $x = 0$ in the distributional sense is equal to δ, as will be proved later.

Keypoint: Distributions are infinitely differentiable, even though their classical counterparts are not differentiable at all.

Some supplementary remarks on the properties 3 and 4 listed above will be given in the discussion below.

8.3.6 Utility of distribution in solving differential equations

Recall that many phenomena studied in physics and engineering are mathematically described using differential equations. A differential equation with respect to a function u is typically written as

$$P(D)u = f, \tag{8.21}$$

where D is a differential operator, $P(D)$ is a polynomial of D, and f is a function. In principle, this class of differential equations can be solved using Fourier transformation. Suppose that we denote the Fourier transform of $f(x)$ and $u(x)$ by $\hat{F}[f](\xi)$ and $\hat{F}[u](\xi)$, respectively,[13] Applying the Fourier transform to both sides of (8.21) gives us

$$\hat{F}[u](\xi) = \frac{\hat{F}[f](\xi)}{P(\xi)}. \tag{8.22}$$

The key point here is that the polynomial $P(D)$ of the operator D is replaced by the polynomial $P(\xi)$ of a variable ξ. Finally, we perform the inverse Fourier transformation of both sides of (8.22) and obtain

$$u(x) = \frac{1}{P(\xi)}\hat{F}^{-1}\left[\hat{F}[f]\right](x), \tag{8.23}$$

which is a solution to the original differential equations.

It should be noted that in order to use this approach, the functions f, u and $P(\xi)^{-1}$ need to be integrable, and this substantial requirement narrows the applicability of

[13]The notation $\hat{F}[f](\xi)$ implies that $\hat{F}[f]$, obtained after the Fourier transformation of f, is a function with respect to the variable ξ. If the variable does not need to be explicit, we say that $\hat{F}[f]$ is the Fourier transform of f.

the approach. The concept of distributions provides a breakthrough in overcoming this obstacle. It makes the functions f, u and $P(\xi)^{-1}$ integrable as distributions, even though they are non-integrable in the classical sense. Hence, the differential equations that involve non-integrable functions (in the classical sense) can be solved using a formulation based on the Fourier transform.

8.4 Theory of distribution

8.4.1 Rapidly decreasing function

Let us examine the practical aspects of distributions. In developing distribution theory, we need only consider a special class of the function $\varphi(x)$, which a distribution maps to the complex numbers. In view of physical and engineering applications, the most important class to consider is that of the rapidly decreasing functions. A function $\varphi(x)$ is rapidly decreasing if it shows a sharp decline when $|x| \to +\infty$.

A typical example of a rapidly decreasing function is the **Gaussian function** given by

$$\varphi(x) = e^{-ax^2}, \quad a > 0. \tag{8.24}$$

In fact, the value of e^{-ax^2} approaches zero very quickly as the distance from $x = 0$ increases. This can be seen in Fig. 8.6.

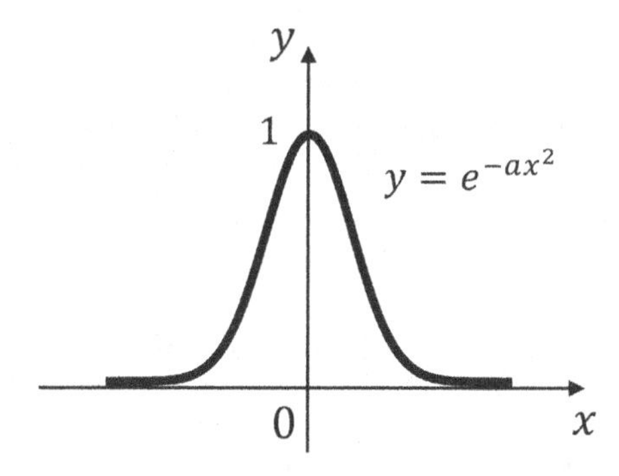

Figure 8.6: A Gaussian function e^{-ax^2}.

An interesting property of Gaussian functions is that their Fourier transforms are also Gaussian functions.[14] This property holds for all rapidly decreasing

[14]This is proved by

$$\frac{1}{\sqrt{2\pi}}\int_{-\infty}^{\infty} e^{-ax^2}e^{-ikx}dx = \frac{1}{\sqrt{2\pi}}\int_{-\infty}^{\infty} e^{-a\left(x+\frac{ik}{2a}\right)^2-\frac{k^2}{4a}}dx = \frac{1}{\sqrt{2\pi a}}e^{-\frac{k^2}{4a}}\int_{-\infty}^{\infty} e^{-u^2}du = \frac{1}{\sqrt{2a}}e^{-\frac{k^2}{4a}},$$

where the variable transformation $u \equiv \sqrt{a}\left(x+\frac{ik}{2a}\right)$ and the well-known formula $\int_{-\infty}^{\infty} e^{-u^2}du = \sqrt{\pi}$ were used.

functions. Therefore, rapidly decreasing functions are compatible with our method for solving differential equations using Fourier transformation. In addition to Gaussian functions, many functions belonging to the Lebesgue space $L^p(\mathbb{R})$ with $1 \le p \le \infty$ are also rapidly decreasing functions.

Let us give an exact definition of rapidly decreasing functions.[15]

Definition (Rapidly decreasing function):
Rapidly decreasing functions $\varphi(x)$ are those satisfying the following conditions:
1. $\varphi(x)$ is infinitely differentiable.
2. For arbitrary non-negative integers α and k, we have[16]

$$\lim_{|x|\to\infty} (1+|x|)^k |\partial^\alpha \varphi(x)| = 0. \quad \blacksquare \tag{8.25}$$

Plainly speaking, condition 2 means that, for large $|x|$, the sequence $\varphi'(x)$, $\varphi''(x)$, $\varphi'''(x)$, ... must decrease sharply. The required rate of decrease is implied by the presence of the factor $(1+|x|)^k$. Note that this expanded polynomial of $|x|$ to degree k is

$$(1+|x|)^k = 1 + c_1|x| + c_2|x|^2 + \cdots + c_{k-1}|x|^{k-1} + |x|^k, \tag{8.26}$$

in which all the terms (except for "1") are monotonically increasing with $|x|$. Hence, to comply with condition 2, $\varphi(x)$ and its derivatives should decrease faster than $|x|^{-N}$, where N is an arbitrary natural number. Of course, the inverse power functions, $|x|^{-3}$, $|x|^{-10}$ and $|x|^{-1000}$, also decrease very quickly with an increase in x. Therefore, $\varphi(x)$ must exceed the decreasing rate of $|x|^{-10000000}$, for example. The same condition applies to all the derivatives $\varphi'(x)$, $\varphi''(x)$, $\varphi'''(x)$, ...

8.4.2 Space of rapidly decreasing functions $S(\mathbb{R})$

Hereafter, we will denote by $S(\mathbb{R})$ a function space composed of the rapidly decreasing functions defined on the real number space.

Example:
If

$$\varphi(x) = e^{-\kappa x^2}, \ \kappa > 0, \tag{8.27}$$

then

$$\varphi \in S(\mathbb{R}). \tag{8.28}$$

[15] A rapidly decreasing function is sometimes referred to as a **sample function** or a **test function**, especially in the field of signal processing.

[16] The symbol ∂^α indicates the α-degree derivative.

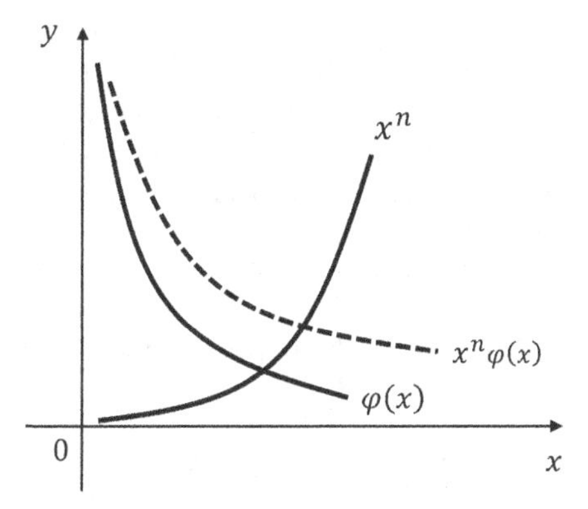

Figure 8.7: Rapidly decreasing function $\varphi(x)$.

Meaning, a Gaussian function is a rapidly decreasing function. ■

The rapidly decreasing property of $\varphi(x) = e^{-\kappa x^2}$ can be proved by satisfying conditions 1 and 2 provided in Definition of rapidly decreasing functions earlier mentioned. First, it is obvious that $\varphi(x) = e^{-\kappa x^2}$ is infinitely differentiable, and condition 1 is satisfied. Using a non-negative integer α, we can write

$$\partial^\alpha \varphi(x) = P_\alpha(x) e^{-\kappa x^2}, \tag{8.29}$$

where $P_\alpha(x)$ is a polynomial of x with degree α. This expression implies

$$|P_\alpha(x)| \leq C_\alpha (1+|x|)^\alpha \tag{8.30}$$

in which C_α is an appropriately chosen positive constant.

In order to examine the property in 2, we combine the above two results and obtain

$$\begin{aligned} (1+|x|)^k |\partial^\alpha \varphi(x)| &= (1+|x|)^k |P_\alpha(x)| e^{-\kappa|x|^2} \\ &\leq C_\alpha (1+|x|)^{\alpha+k} e^{-\kappa|x|^2}, \end{aligned} \tag{8.31}$$

which holds for arbitrary non-positive integers α and k. Furthermore, it trivially follows that[17]

$$\lim_{|x|\to\infty} \frac{(1+|x|)^{\alpha+k}}{e^{\kappa x^2}} = 0. \tag{8.32}$$

Therefore, we conclude that

$$\lim_{|x|\to\infty} (1+|x|)^k |\partial^\alpha \varphi(x)| = 0. \tag{8.33}$$

[17] The rate of increase for e^{x^2} is extremely large. It exceeds the rate of increase for power functions $|x|^n$ for any large n.

In order to further the argument, we need to define the meaning of convergence in the sense of $S(\mathbb{R})$. Recall that, in general, different functional spaces are endowed with different definitions of convergence (or distance between elements). For example, the convergence of a sequence of functions in $L^p(\mathbb{R})$ is completely different[18] from that in $C(\mathbb{R})$. Similarly, in $S(\mathbb{R})$, a special convergence definition should be given for the sequence of functions $\{\varphi_j\} \subset S(\mathbb{R})$.

Definition (Convergence in $S(\mathbb{R})$):
Suppose that an infinite sequence of functions $\{\varphi_j\} \subset S(\mathbb{R})$ converges to φ in $S(\mathbb{R})$. Then, for arbitrary non-positive integers k and α, we have

$$\lim_{j\to\infty} \|\varphi_j - \varphi\|_{k,\alpha} = 0. \tag{8.34}$$

Here, the symbol $\|\cdots\|_{k,\alpha}$ is the norm in $S(\mathbb{R})$, which is defined by[19]

$$\|\varphi\|_{k,\alpha} = \sup (1+|x|)^k |\partial^\alpha \varphi(x)|, \ (-\infty < x < \infty) \tag{8.35}$$

We express this convergence of φ_j to φ in S by

$$\varphi_j \to \varphi \text{ in } S \ (j\to\infty). \quad \blacksquare \tag{8.36}$$

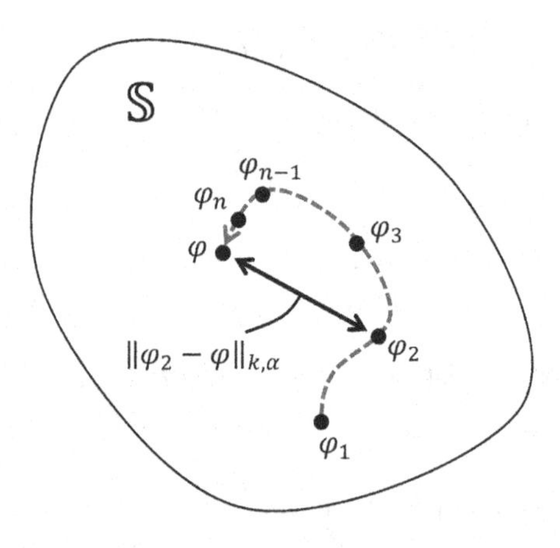

Figure 8.8: The norm $\|\varphi_2 - \varphi\|_{k,\alpha}$ in the sense of S.

It should be noted that, in (8.35), because φ is rapidly decreasing, $(1+|x|)^k |\partial^\alpha \varphi(x)|$ converges to zero as $|x| \to \infty$. However, for a finite x, we can show that the term $(1+|x|)^k |\partial^\alpha \varphi(x)|$ has a finite value. Its value alters depending on the choices

[18] In the space $L^p(\mathbb{R})$, the distance $\|f-g\|_p$ between the functions $f, g \in L^p(\mathbb{R})$ is defined using the Lebesgue integral, by $\|f-g\|_p = \int |f-g|^p d\mu$; See section 6.5.1.

[19] The value of the norm $\|\varphi\|_{k,\alpha}$ depends on the choices of k and α.

of k and α, even for a specific x. Therefore, we search for the supremum, $\sup(1+|x|)^k |\partial^\alpha \varphi(x)|$, for given k and α by scanning the x axis from $-\infty$ to ∞. The obtained supremum is what we call the norm of φ in $S(\mathbb{R})$ for the fixed k and α. Certainly, different values of k and α lead to different values of $||\varphi||_{k,\alpha}$. Therefore, if we want to prove

$$\varphi_j \to \varphi \text{ in } S \ (j \to \infty), \tag{8.37}$$

then we need to verify

$$\lim_{j\to\infty} ||\varphi_j - \varphi||_{k,\alpha} = 0 \tag{8.38}$$

for arbitrary choices of k and α.

Keypoint: The norm $||\cdots||_{k,\alpha}$ in the sense of $S(\mathbb{R})$ is fundamentally different from the norms in other functional spaces, such as $||\cdots||_p$ in $L^p(\mathbb{R})$.

8.4.3 Tempered distribution

This section highlights a special class of distributions T whose definitions are based on rapidly decreasing functions $\varphi \in S(\mathbb{R})$. The $S(\mathbb{R})$-based distribution, which maps $\varphi \in S(\mathbb{R})$ to $\alpha \in \mathbb{C}$, is called a tempered distribution. After providing its precise definition, section 8.4.4 will explain why these distributions are "tempered."

Definition (Tempered distribution):
Suppose a mapping $T : S(\mathbb{R}) \ni \varphi \to \alpha \in \mathbb{C}$ satisfies the following two conditions, then we call T a **tempered distribution**.

1. (Linearity) For arbitrary $\varphi, \phi \in S(\mathbb{R})$ and $\alpha, \beta \in \mathbb{C}$, we have

$$\langle T, \alpha\varphi + \beta\phi \rangle = \alpha\langle T, \varphi\rangle + \beta\langle T, \phi\rangle. \tag{8.39}$$

2. (Continuity) If $\varphi_j \to \varphi$ $(j \to \infty)$ in $S(\mathbb{R})$, we have

$$\langle T, \varphi_j\rangle \to \langle T, \varphi\rangle \ (j \to \infty). \quad \blacksquare \tag{8.40}$$

Henceforth, we will denote the set of all tempered distributions by $S'(\mathbb{R})$.

8.4.4 Distribution defined by integrable function

When creating a distribution, we often use an integrable function $f(x)$, as in the following definition.

$$T_f : \varphi \to \langle T_f, \varphi\rangle = \int_{-\infty}^{\infty} f(x)\varphi(x)dx, \quad [\varphi \in S(\mathbb{R})]. \tag{8.41}$$

Here, the subscript f in T_f means that the distribution T_f provides a number $[\int_{-\infty}^{\infty} f(x)\varphi(x)dx]$ whose value is determined by integration involving $f(x)$. This distribution class T_f, whose definition is given by the integral in (8.41), is called a **regular distribution**. Otherwise, if a distribution is not defined by (8.41), it is called an **anomalous distribution**. The following three earlier mentioned distributions

$$\langle \delta, \varphi \rangle,\ \langle H, \varphi \rangle \text{ and} \left\langle \mathrm{PV}\frac{1}{x}, \varphi \right\rangle \tag{8.42}$$

are all anomalous distributions.

The root of the word "tempered" comes from the integral-based definition of (8.41). In this definition, the product of $f(x)$ and $\varphi(x)$ appears in the integrand. In order for the integral to exist, $f(x)$ cannot increase sharply. If $f(x)$ shows a steep increase with $|x|$, then the product $f(x)\varphi(x)$ diverges as $|x| \to \infty$. Therefore, the growth of $f(x)$ should be repressed by the decreasing behavior of $\varphi(x)$. This means that the rate of increase of $f(x)$ must be tempered enough to let (8.41) make sense. This is the reason why the $S(\mathbb{R})$-based regular distribution, which maps $\varphi(x)$ to a complex number as exemplified by (8.41), is called a tempered distribution. The same reasoning applies to the $S(\mathbb{R})$-based anomalous distributions, if their mapping natures are described in a similar expression to (8.41); see (8.56).

Incidentally, a broad class of functions $f(x)$ can be used to define the regular distribution T_f. Broadly speaking, the integrability of these functions suffices. Even when $f(x)$ involves discontinuity or divergence, it is sufficient that it is integrable in the sense of $L^p(\mathbb{R})$ (see Example 1 just below). This wide acceptable range for $f(x)$ is one of the merits of regular distributions. This advantage is seen in the differentiation and Fourier transformation of many ill-defined functions.

Example 1:
Suppose that $f \in L^p(\mathbb{R})$ for $1 \le p \le \infty$, then T_f defined by (8.41) is a tempered distribution. ■

Proof:
Linearity in T_f is trivial, so we will establish only the continuity of T_f in two different cases: $p = 1$ and $1 < p \le \infty$.

1. Set $p = 1$, and suppose the sequence φ_j is such that $\varphi_j \to \varphi$ in S as $j \to \infty$. It follows that

$$\begin{aligned} \left|\langle T_f, \varphi_j \rangle - \langle T_f, \varphi \rangle\right| &\le \int_{-\infty}^{\infty} |f(x)| \cdot \left|\varphi_j(x) - \varphi(x)\right| dx \\ &\le \left\|\varphi_j - \varphi\right\|_{0,0} \int_{-\infty}^{\infty} |f(x)| dx \to 0 \text{ in } S \ (j \to \infty). \end{aligned} \tag{8.43}$$

This implies the continuity of T_f:

$$\langle T_f, \varphi_j \rangle \to \langle T_f, \varphi \rangle \text{ in } S \quad (j \to \infty). \tag{8.44}$$

2. Set $1 < p \leq \infty$. It follows from **Hölder's inequality**[20] that

$$\left|\langle T_f, \varphi_j - \varphi\rangle\right| \leq \|f\|_{L^p} \cdot \left\|\varphi_j - \varphi\right\|_{L^q} \quad \left(\frac{1}{p} + \frac{1}{q} = 1\right). \tag{8.45}$$

In addition, we see from the definition of L^p-norm that

$$\begin{aligned} \left\|\varphi_j - \varphi\right\|_{L^q} &= \left[\int_{-\infty}^{\infty} \frac{1}{\left(1+|x|^2\right)^q} \cdot \left\{\left(1+|x|^2\right) \cdot \left|\varphi_j(x) - \varphi(x)\right|\right\}^q dx\right]^{1/q} \\ &\leq \left[C\left(\left\|\varphi_j - \varphi\right\|_{0,0} + \left\|\varphi_j - \varphi\right\|_{2,0}\right)^q\right]^{1/q}, \end{aligned} \tag{8.46}$$

where

$$C = \int_{-\infty}^{\infty} \frac{1}{\left(1+|x|^2\right)^q} dx < \infty. \tag{8.47}$$

Combining these results, we obtain

$$\left|\langle T_f, \varphi_j\rangle - \langle T_f, \varphi\rangle\right| \leq \|f\|_{L^p} \cdot C^{\frac{1}{q}} \left(\left\|\varphi_j - \varphi\right\|_{0,0} + \left\|\varphi_j - \varphi\right\|_{2,0}\right) \to 0 \text{ in } S \; (j \to \infty), \tag{8.48}$$

which implies that

$$\langle T_f, \varphi_j\rangle \to \langle T_f, \varphi\rangle \text{ in } S \; (j \to \infty). \quad \blacksquare \tag{8.49}$$

Example 2:
Suppose a continuous function $f(x)$ satisfies the inequality

$$|f(x)| \leq c(1+|x|)^N \; (x \in \mathbb{R}). \tag{8.50}$$

Here, c is a positive constant, and N is a natural number. Under these conditions, the distribution T_f, characterized by $f(x)$, is a tempered distribution.[21] ■

Proof:
The linearity of T_f is obvious, and the continuity of T_f is confirmed by the following calculation:

$$\begin{aligned} &\left|\langle T_f, \varphi_j\rangle - \langle T_f, \varphi\rangle\right| \\ \leq\; & \int_{-\infty}^{\infty} c(1+|x|)^N \left|\varphi_j(x) - \varphi(x)\right| dx \\ =\; & \int_{-\infty}^{\infty} c(1+|x|)^N (1+|x|)^{-N-2} \times (1+|x|)^{N+2} \left|\varphi_j(x) - \varphi(x)\right| dx \\ \leq\; & c\left[\int_{-\infty}^{\infty} \frac{dx}{(1+|x|)^2}\right] \times \left\|\varphi_j - \varphi\right\|_{N+2,0} \to 0 \text{ in } S \; (j \to \infty). \quad \blacksquare \end{aligned} \tag{8.51}$$

---

[20]Hölder's inequality is expressed by $\|fg\|_{L^1} \leq \|f\|_{L^p} \cdot \|g\|_{L^q}$, where p and q are real and satisfy $1 \leq p < \infty$, $1 \leq q < \infty$, and $\frac{1}{p} + \frac{1}{q} = 1$. Specifically, when $p = q = 2$, it is equivalent to the well-known **Cauchy-Schwarz inequality**: $|\int f(x)g(x)^* dx| \leq \left[\int |f(x)|^2 dx\right]^{1/2} \cdot \left[\int |g(x)|^2 dx\right]^{1/2}$.

[21]In this case, $f(x)$ is called a tempered continuous function.

8.4.5 Identity between function and distribution

Thus far, we introduced two families of distributions. One consists of regular distributions T_f, which are characterized by integrable functions $f(x)$:

$$T_f : S \ni \varphi \to \langle T_f, \varphi \rangle = \int_{-\infty}^{\infty} f(x)\varphi(x)dx. \tag{8.52}$$

In this case, the number $\langle T_f, \varphi \rangle$ is determined by an integral involving $f(x)$ in its integrand. The other is composed of anomalous distributions T:

$$T : S \ni \varphi \to \langle T, \varphi \rangle. \tag{8.53}$$

Here, no universal prescription exists for evaluating the number $\langle T, \varphi \rangle$, and the computation of this number depends on the nature of the distribution.

The essential feature of a regular distribution T_f, characterized by an integrable function $f(x)$, is the one-to-one correspondence between T_f and f. Namely, the nature of T_f is uniquely determined by $f(x)$; this is written by

$$T_f = T_g \Longleftrightarrow f(x) = g(x). \tag{8.54}$$

This one-to-one correspondence leads us to an important consequence: defining T_f is equivalent to defining $f(x)$. In other words, mathematical manipulations, such as differentiation or Fourier transformation, on $f(x)$ are identified with those on the distribution T_f, which is characterized by $f(x)$. In this context, defining the function $f(x)=e^{-\kappa|x|^2}$ is the same as defining the mapping

$$T_f : \varphi \to \int_{-\infty}^{\infty} e^{-\kappa|x|^2} \varphi(x)dx. \tag{8.55}$$

A similar discussion holds true for anomalous distributions. The only difference from the regular distribution case is the absence of the characteristic function $f(x)$. In fact, no function corresponds to the anomalous distribution T. By necessity, therefore, we introduce a sham function $T(x)$ that satisfies the following relation:

$$T : \varphi \to \langle T, \varphi \rangle = \int_{-\infty}^{\infty} T(x)\varphi(x)dx. \tag{8.56}$$

Remember that $T(x)$ is not a function in the classical sense, and this integration does not make sense. But if so, why did we introduce such a superficial mathematical entity? The reason is the utility of the notation $T(x)$ in computing the differentiation and Fourier transformation in the distributional sense. These merits will be demonstrated in the next two sections.

8.5 Mathematical manipulation of distribution

8.5.1 Distributional derivative

In this section, we develop differentiation for tempered distributions.

Definition (Distributional derivative):
For a regular distribution T_f, characterized by a function f, we introduce a mapping, denoted by[22] $\partial^\alpha T_f$:

$$\partial^\alpha T_f: \ \varphi \to \langle T_{\partial^\alpha f}, \varphi \rangle = \int_{-\infty}^{\infty} \partial^\alpha f(x) \varphi(x) dx. \tag{8.57}$$

This mapping is called the **distributional derivative**[23] of T_f. ■

Remember that it is possible to differentiate $\varphi \in S(\mathbb{R})$ as many times as we would like and that φ is rapidly decreasing, which means $\varphi(x) \to 0$ as $|x| \to \infty$. These facts imply the following partial integration.

$$\langle T_{\partial^\alpha f}, \varphi \rangle = \int_{-\infty}^{\infty} \partial^\alpha f(x) \varphi(x) dx = (-1)^\alpha \int_{-\infty}^{\infty} f(x) \partial^\alpha \varphi(x) dx = \langle T_f, (-1)^\alpha \partial^\alpha \varphi \rangle \tag{8.58}$$

The result indicates that the distributional derivative $\partial^\alpha T_f$ is a mapping from φ to the number $\langle T_f, (-1)^\alpha \partial^\alpha \varphi \rangle$.

Theorem (Derivative of a regular distribution):
The derivative $\partial^\alpha T_f$ of a regular distribution T_f is a mapping expressed by

$$\partial^\alpha T_f : \varphi \to \langle T_f, (-1)^\alpha \partial^\alpha \varphi \rangle. \quad ■ \tag{8.59}$$

It is worth noting in (8.59), the mathematical operation of differentiation is imposed onto the function $\varphi(x)$. Because T_f is not a function and, thus, not able to be differentiated in a usual sense, we make $\varphi(x)$ a substitute. Namely, we differentiate $\varphi(x)$ instead of T_f, by which we claim to obtain the derivative of T_f.

The same story applies to anomalous distributions.

Theorem (Derivative of an anomalous distribution):
For a given anomalous distribution T, its distributional derivative $\partial^\alpha T$ is expressed by the mapping

$$\partial^\alpha T: \ \varphi \to \langle T, (-1)^\alpha \partial^\alpha \varphi \rangle. \quad ■ \tag{8.60}$$

---

[22] The symbol ∂^α in $\partial^\alpha T_f$ needs careful consideration. It is a dummy operator, which does not imply the differentiation of a given function. Remember that T_f is not a function, so $\partial^\alpha T_f$ must not represent a derivative of T_f in a usual sense.

[23] It is also called a **weak derivative** or a **derivative in the sense of the distribution** .

In this case, no relevant integration exists that can evaluate the number to which φ is mapped. The number is to be evaluated according to the nature of T.

The following examples apply the distributional derivative to sham functions, which are non-differentiable in the classical sense.

Example 1:
Consider the distributional derivative of Heaviside's step function H, which we conventionally denote as ∂H. From the definition of a distributional derivative given by (8.60), we have

$$\partial H: \ \varphi \to \langle H, -\partial\varphi \rangle. \tag{8.61}$$

It follows that

$$\langle H, -\partial\varphi \rangle = -\int_{-\infty}^{\infty} H(x)\varphi'(x)dx = -\int_0^{\infty} \varphi'(x)dx = \varphi(0) = \langle \delta, \varphi \rangle, \tag{8.62}$$

which means

$$\partial H: \ \varphi \to \langle \delta, \varphi \rangle. \tag{8.63}$$

Thus, we have the well-known result[24] that states the derivative of H is Dirac's delta function δ. ■

In physics and engineering, this result may seem trivial. In these cases, H and δ are often mathematically manipulated as though they were functions, and the following deceptive relation is conventionally used:

$$\frac{d}{dx}H(x) = \delta(x). \tag{8.64}$$

This relation is meaningless in the usual sense but is justified if we translate it into a distributional argument.

Example 2:
Suppose we have an infinite series of functions defined by

$$f(x) = \sum_{n=1}^{\infty} \frac{\sin nx}{n}. \tag{8.65}$$

Clearly, $f(x)$ is a periodic function of x with a period of 2π. In the range $[-\pi, \pi]$, it is explicitly written as

$$f(x) = \begin{cases} -\dfrac{x}{2} + \dfrac{\pi}{2} & 0 < x \le \pi \\ 0 & x = 0 \\ -\dfrac{x}{2} - \dfrac{\pi}{2} & -\pi \le x < 0 \end{cases} \tag{8.66}$$

[24]This result is justified only when H and δ are regarded as distributions, not functions.

Figure 8.9 illustrates the x-dependence of $f(x)$. Due to the discontinuous jumps at $x=0,\ \pm 2\pi,\ \pm 4\pi,\cdots$, it is impossible to differentiate $f(x)$ at those points. However, differentiation becomes possible when we replace $f(x)$ by T_f, the regular distribution characterized by f. After a straightforward calculation, we produce the distributional derivative of T_f:

$$-\frac{1}{2}+\pi\sum_{k=-\infty}^{\infty}\delta_{2k\pi}, \tag{8.67}$$

which satisfies the relation:[25]

$$\langle T_{\partial f},\varphi\rangle=\langle T_f,-\partial\varphi\rangle=-\int_{-\infty}^{\infty}f(x)\frac{d\varphi(x)}{dx}dx=\left\langle -\frac{1}{2}+\pi\sum_{k=-\infty}^{\infty}\delta_{2k\pi},\ \varphi\right\rangle. \quad \blacksquare \tag{8.69}$$

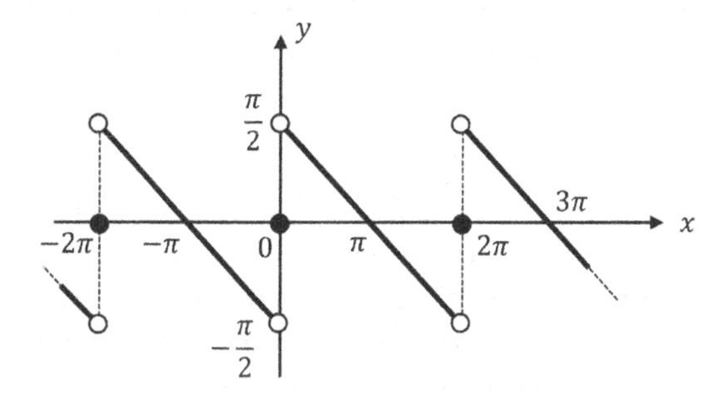

Figure 8.9: Profile of $y=f(x)$ given by (8.66).

It is helpful to return to the original infinite series of (8.65) and term-wise derive it. Because

$$\left(\frac{\sin nx}{n}\right)'=\cos nx, \tag{8.70}$$

---

[25]The derivation of (8.67) is shown below:

$$\begin{aligned}
\langle T_{\partial f},\varphi\rangle &= \langle T_f,-\partial\varphi\rangle = -\int_{-\infty}^{\infty}f(x)\frac{d\varphi(x)}{dx}dx \\
&= -\sum_{k=-\infty}^{\infty}\left[\int_{(2k-1)\pi}^{2k\pi}\left(-\frac{x}{2}-\frac{\pi}{2}\right)\frac{d\varphi(x)}{dx}dx+\int_{2k\pi}^{(2k+1)\pi}\left(-\frac{x}{2}+\frac{\pi}{2}\right)\frac{d\varphi(x)}{dx}dx\right] \\
&= -\sum_{k=-\infty}^{\infty}\left[-\frac{\pi}{2}\varphi(2k\pi)+\frac{1}{2}\int_{(2k-1)\pi}^{2k\pi}\varphi(x)dx-\frac{\pi}{2}\varphi(2k\pi)+\frac{1}{2}\int_{2k\pi}^{(2k+1)\pi}\varphi(x)dx\right] \\
&= \sum_{k=-\infty}^{\infty}\left[\pi\varphi(2k\pi)-\frac{1}{2}\int_{(2k-1)\pi}^{(2k+1)\pi}\varphi(x)dx\right] = \int_{-\infty}^{\infty}\left[-\frac{1}{2}+\pi\sum_{k=-\infty}^{\infty}\delta(x-2k\pi)\right]\varphi(x)dx \\
&= \left\langle -\frac{1}{2}+\pi\sum_{k=-\infty}^{\infty}\delta_{2k\pi},\ \varphi\right\rangle. \quad \blacksquare
\end{aligned} \tag{8.68}$$

we have

$$f'(x) = \sum_{n=1}^{\infty} \cos(nx). \tag{8.71}$$

The infinite sum to the right diverges at $x = 2k\pi$, which is why the derivative of $f(x)$ does not exist in the usual sense. We have already seen that this problem is overcome by replacing it with the corresponding distribution. The concept of the distribution reactivates the mathematical significance of diverging infinite series. This advantage holds true for many kinds of infinite series and has been primarily used in quantum field theory and other branches of physics.

8.5.2 Fourier transformation using distribution

Now, we are ready to introduce the concept of Fourier transformation in the distributional sense. Originally, the Fourier transformation of a function $f(x)$ performs a special class of integration defined by

$$\hat{F}[f](\xi) = \frac{1}{\sqrt{2\pi}} \int_{-\infty}^{\infty} f(x) e^{-i\xi x} dx. \tag{8.72}$$

Integration allows us to obtain a new function with respect to ξ, which we denote by $\hat{F}[f](\xi)$. However, this integral does not always exist. For example, it will diverge when $f(x)$ monotonically increases with x or even when $f(x)$ is constant up to infinity. This problem can also be overcome by introducing the concept of Fourier transformation in the sense of a distribution.

Definition (Fourier transform of distribution):
The mapping

$$\hat{F}\left[T_f\right] : \varphi \to \langle T_{\hat{F}[f]}, \varphi \rangle \tag{8.73}$$

is called the Fourier transform $\hat{F}\left[T_f\right]$ of the distribution T_f. ■

We should note that both $\hat{F}\left[T_f\right]$ and T_f are mappings from φ to a number. Therefore, the Fourier transform of T_f is a transformation between mappings from T_f to $\hat{F}[T_f]$, namely, from

$$T_f : \varphi \to \langle T_f, \varphi \rangle = \int_{-\infty}^{\infty} f(x)\varphi(x) dx \tag{8.74}$$

to

$$\hat{F}\left[T_f\right] : \varphi \to \langle T_{\hat{F}[f]}, \varphi \rangle = \int_{-\infty}^{\infty} \hat{F}[f](\xi)\varphi(\xi) d\xi. \tag{8.75}$$

Let us examine how to compute the number $\langle T_{\hat{F}[f]}, \varphi \rangle$ provided by the Fourier transformation. In order to do this, we employ the following formula[26] with respect

[26] The formula for this is derived as follows:

to functions $f, g \in S(\mathbb{R})$ and their respective Fourier transforms $\hat{F}[f]$ and $\hat{F}[g]$:

$$\int_{-\infty}^{\infty} \hat{F}[f](x) \cdot g(x) dx = \int_{-\infty}^{\infty} f(x) \cdot \hat{F}[g](x) dx. \tag{8.76}$$

Using this relation, the explicit form of $\langle T_{\hat{F}[f]}, \varphi \rangle$ is rewritten as

$$\begin{aligned} \langle T_{\hat{F}[f]}, \varphi \rangle &= \int_{-\infty}^{\infty} \hat{F}[f](x)\, \varphi(x) dx = \int_{-\infty}^{\infty} f(x) \hat{F}[\varphi](x) dx \\ &= \langle T_f, \hat{F}[\varphi] \rangle. \end{aligned} \tag{8.77}$$

The last term is the result of the Fourier transform $\hat{F}\left[T_f\right]$. In summary, the Fourier transform $\hat{F}\left[T_f\right]$ of the distribution T_f is equivalent to a mapping represented by[27]

$$\varphi \to \langle T_f, \hat{F}[\varphi] \rangle. \tag{8.78}$$

In (8.78), it should be noted that φ is again used as a substitute. In order to evaluate the Fourier transform of T_f, we imposed the transformation on φ instead of on T_f, similar to the distributional differentiation case; See (8.59). We know that T_f is not a function, but φ must be a function. Therefore, the Fourier transformation is operated on φ and not on T_f, which was how we defined the Fourier transform of T_f.

Example 1:
Assume $f(x) = c$ is a constant function, and let us find the Fourier transform of T_f characterized by f. From the definition, we have

$$\begin{aligned} \hat{F}\left[T_f\right] : \varphi \to \langle T_{\hat{F}[f]}, \varphi \rangle &= \langle T_f, \hat{F}[\varphi] \rangle = \int_{-\infty}^{\infty} c\hat{F}[\varphi](x) dx \\ &= \sqrt{2\pi}c \cdot \frac{1}{\sqrt{2\pi}} \int_{-\infty}^{\infty} e^{i \times 0 \times x} \hat{F}[\varphi](x) dx \\ &= \sqrt{2\pi}c \cdot \hat{F}^{-1}\left[\hat{F}[\varphi]\right](0) = \sqrt{2\pi}c \cdot \varphi(0) \\ &= \langle \sqrt{2\pi}c\delta, \varphi \rangle. \end{aligned} \tag{8.79}$$

This leads us to conclude that

$$\hat{F}\left[T_f\right] : \varphi \to \langle \sqrt{2\pi}c\delta, \varphi \rangle. \tag{8.80}$$

$\int_{-\infty}^{\infty} \hat{F}[f](\xi) \cdot g(\xi) e^{ix\xi} d\xi = \int_{-\infty}^{\infty} \left[\frac{1}{\sqrt{2\pi}} \int_{-\infty}^{\infty} f(y) e^{-i\xi y} dy\right] \cdot g(\xi) e^{ix\xi} d\xi$

$= \int_{-\infty}^{\infty} \left[\frac{1}{\sqrt{2\pi}} \int_{-\infty}^{\infty} f(y) e^{-i(y-x)\xi} dy\right] \cdot g(\xi) d\xi = \int_{-\infty}^{\infty} \left[\frac{1}{\sqrt{2\pi}} \int_{-\infty}^{\infty} f(x+z) e^{-iz\xi} dz\right] \cdot g(\xi) d\xi$

$= \int_{-\infty}^{\infty} f(x+z) \left[\frac{1}{\sqrt{2\pi}} \int_{-\infty}^{\infty} g(\xi) e^{-iz\xi} d\xi\right] dz = \int_{-\infty}^{\infty} f(x+z) \hat{F}[g](z) dz.$

To obtain the fourth term from the third, the variable y was replaced with $z \equiv y - x$. Imposing $x = 0$ onto the first and last terms, we produce the desired formula: $\int_{-\infty}^{\infty} \hat{F}[f](\xi) \cdot g(\xi) d\xi = \int_{-\infty}^{\infty} f(z) \hat{F}[g](z) dz$.

[27] The same argument results in the inverse Fourier transform of T_f, designated by $\hat{F}^{-1}\left[T_f\right]$, being written as $\hat{F}^{-1}\left[T_f\right] : \varphi \to \langle T_{\hat{F}^{-1}[f]}, \varphi \rangle = \langle T_f, \hat{F}^{-1}[\varphi] \rangle$, $(\varphi \in S(\mathbb{R}))$.

Consequently, the Fourier transform of a constant function (in the sense of a distribution) is proportional to Dirac's delta function. Note that both T_f and $\sqrt{2\pi}c\delta$ are distributions (or mappings), not functions. ■

These results agree with the expedience formulae that frequently appear in physics and engineering:

$$\frac{1}{\sqrt{2\pi}}\int_{-\infty}^{\infty} ce^{-i\xi x}dx \propto \delta(x),$$
$$\frac{1}{\sqrt{2\pi}}\int_{-\infty}^{\infty} \delta(x)e^{-i\xi x}dx = \text{const.} \tag{8.81}$$

Both formulae are improper due to their diverging properties. However, essentially identical results are obtained through distributional computation.

Example 2:
Suppose we have a function $f(x) = e^{iax}$ and its corresponding distribution T_f. Its Fourier transform $\hat{F}\left[T_f\right]$ is given by

$$\begin{aligned}\hat{F}\left[T_f\right]: \varphi \to \langle T_{\hat{F}[f]}, \varphi\rangle &= \langle T_f, \hat{F}[\varphi]\rangle = \int_{-\infty}^{\infty} e^{iax}\hat{F}\left[\varphi\right](x)dx \\ &= \sqrt{2\pi}\cdot\frac{1}{\sqrt{2\pi}}\int_{-\infty}^{\infty} e^{iax}\hat{F}\left[\varphi\right](x)dx \\ &= \sqrt{2\pi}\cdot\hat{F}^{-1}\left[\hat{F}\left[\varphi\right]\right](a) \\ &= \sqrt{2\pi}\cdot\varphi\left(a\right) = \langle\sqrt{2\pi}\delta_a, \varphi\rangle,\end{aligned} \tag{8.82}$$

and we can conclude that

$$\hat{F}\left[T_f\right]: \varphi \to \langle\sqrt{2\pi}\delta_a, \varphi\rangle. \tag{8.83}$$

The Fourier transform of $f(x) = e^{iax}$ is the product of Dirac's delta δ_a and $\sqrt{2\pi}$. Using the language of sham functions, we can write it as

$$\frac{1}{\sqrt{2\pi}}\int_{-\infty}^{\infty} e^{iax}e^{-i\xi x}dx = \sqrt{2\pi}\delta\left(\xi - a\right). \quad ■ \tag{8.84}$$

Example 3:
Consider the power function $f(x) = x^n$. It follows that

$$\hat{F}\left[T_f\right]: \varphi \to \langle T_{\hat{F}[f]}, \varphi\rangle = \langle T_f, \hat{F}[\varphi]\rangle = \int_{-\infty}^{\infty} x^n\hat{F}\left[\varphi\right](x)dx. \tag{8.85}$$

Partial integration with n repetitions yields

$$\begin{aligned}x^n\hat{F}\left[\varphi\right](x) &= x^n\cdot\frac{1}{\sqrt{2\pi}}\int_{-\infty}^{\infty} e^{-i\xi x}\varphi(\xi)d\xi = \frac{1}{\sqrt{2\pi}}\int_{-\infty}^{\infty} i^n\frac{d^n e^{-i\xi x}}{d\xi^n}\varphi(\xi)d\xi \\ &= \frac{1}{\sqrt{2\pi}}\int_{-\infty}^{\infty}(-i)^n e^{-i\xi x}\frac{d^n\varphi(\xi)}{d\xi^n}d\xi = (-i)^n\hat{F}\left[\partial^n\varphi\right](x).\end{aligned} \tag{8.86}$$

Integrating the first and last terms with respect to x, we have

$$\begin{aligned}\int_{-\infty}^{\infty} x^n \hat{F}[\varphi](x)dx &= (-i)^n \int_{-\infty}^{\infty} \hat{F}[\partial^n \varphi](x)dx \\ &= (-i)^n \sqrt{2\pi}\cdot\frac{1}{\sqrt{2\pi}}\int_{-\infty}^{\infty} e^{i\times 0\times x}\hat{F}[\partial^n\varphi](x)dx \\ &= (-i)^n \sqrt{2\pi}\cdot\hat{F}^{-1}\left[\hat{F}[\partial^n\varphi]\right](0) \\ &= (-i)^n \sqrt{2\pi}\cdot\left.\frac{d^n\varphi(\eta)}{d\eta^n}\right|_{\eta=0}. \end{aligned} \tag{8.87}$$

The derivative in the last term can be rewritten as

$$\left.\frac{d^n\varphi(\eta)}{d\eta^n}\right|_{\eta=0} = \langle\delta,\partial^n\varphi\rangle = (-1)^n\langle\partial^n\delta,\varphi\rangle. \tag{8.88}$$

Substituting this into (8.87) and comparing the result with (8.85), we conclude that

$$\hat{F}\left[T_f\right] : \varphi \to \langle T_{\hat{F}[f]},\varphi\rangle = i^n\sqrt{2\pi}\langle\partial^n\delta,\varphi\rangle. \tag{8.89}$$

In short, the Fourier transform of $f(x) = x^n$ is proportional to $\partial^n\delta$, the nth derivative of Dirac's delta function. This is conventionally expressed as

$$\frac{1}{\sqrt{2\pi}}\int_{-\infty}^{\infty} x^n e^{-i\xi x}dx = i^n\sqrt{2\pi}\cdot\frac{d^n\delta(\xi)}{d\xi^n}. \quad \blacksquare \tag{8.90}$$

In summary, the Fourier transformation of the three classes of regular distributions mentioned above are:

$$\begin{aligned} f(x) = c &\quad\Leftrightarrow\quad \hat{F}[f] = \sqrt{2\pi}c\cdot\delta \\ f(x) = e^{iax} &\quad\Leftrightarrow\quad \hat{F}[f] = \sqrt{2\pi}\cdot\delta_a \\ f(x) = x^n &\quad\Leftrightarrow\quad \hat{F}[f] = i^n\sqrt{2\pi}\cdot\partial^n\delta \end{aligned} \tag{8.91}$$

Before closing this section, it would be pedagogical to recall the identity relation between the two classes of Fourier transforms: that of the usual sense and that of the distributional sense. The distributional concept is a natural expansion of the usual one, as was shown with Dirac's delta function δ. The same is true for regular distributions. For instance, a function $f \in L^2(\mathbb{R})$ gives both classes of Fourier transforms: $\hat{F}[f]$ and $\hat{F}\left[T_f\right]$, in which the former is a function and the latter is a mapping. The two mathematical entities are computationally identical. This is confirmed by comparing the following two computations. First, we define $\hat{F}[f]$ and then calculate the corresponding distribution $T_{\hat{F}[f]}$. Second, we define T_f and then perform its Fourier transformation $\hat{F}\left[T_f\right]$. Both distributions, $T_{\hat{F}[f]}$ and $\hat{F}\left[T_f\right]$, yield the same number when mapped from φ.

8.5.3 Weak solution to differential equation

Recall that distributions are infinitely differentiable. In physics and engineering, this property is widely used to solve differential equations, which model multiple types of actual phenomena that are beyond the scope of elementary analysis.

In order to understand the basic idea of solving differential equations using distributions, we begin with the linear differential equation represented by

$$\mathscr{L} f(t) = 0, \; -\infty < t < \infty. \tag{8.92}$$

Here, $\mathscr{L}$ is a linear differential operator, such as $\mathscr{L} = d^2/dt^2$. Next, we use the distributional language and assume that the following relation holds for $\varphi(t) \in S(\mathbb{R})$:

$$\langle \mathscr{L} f, \varphi \rangle = \langle 0, \varphi \rangle. \tag{8.93}$$

The right-hand side of this equation is always zero. From this, we derive

$$\langle \mathscr{L} f, \varphi \rangle = \left\langle \frac{d^2 f}{dt^2}, \varphi \right\rangle = \left\langle f, \frac{d^2 \varphi}{dt^2} \right\rangle = \langle f, \mathscr{L} \varphi \rangle = 0. \tag{8.94}$$

This is our key equation. Suppose no solution exists for the original equation $\mathscr{L} f = 0$, but there is an appropriate f that satisfies the relation

$$\langle f, \mathscr{L} \varphi \rangle = 0. \tag{8.95}$$

In this case, we say f is a **weak solution**[28] of the differential equation $\mathscr{L} f = 0$. This concept of a weak solution makes it possible to include discontinuous functions as solutions to differential equations. A simple example of this follows.

Example:
Consider the one-dimensional partial differential equation

$$\frac{\partial^2 w(x,t)}{\partial t^2} = \frac{\partial^2 w(x,t)}{\partial x^2} \tag{8.96}$$

under the following initial conditions:

$$w(x,0) = f(x), \quad \left. \frac{\partial w(x,t)}{\partial t} \right|_{t=0} = g(x). \tag{8.97}$$

Here, $f(x)$ and $g(x)$ are smooth, continuous functions. Equation (8.96) is known to have a smooth, continuous solution described by[29]

$$w(x,t) = \frac{f(x+t) + f(x-t)}{2} + \frac{1}{2} \int_{x-t}^{x+t} g(\xi) d\xi. \tag{8.98}$$

[28] In some texts, it is called a **distributional solution** or a **solution in the sense of distribution**.

[29] This is called a **d'Alembert's solution** or **d'Alembert's formula**, broadly known in physics and engineering.

In addition, it has a weak solution, which can be written as

$$w(x,t) = c\left[1 - H(x-t)\right], \tag{8.99}$$

where $H(x)$ is Heaviside's step function and c is a non-zero constant. This discontinuous solution corresponds to a shock wave[30] and is an acceptable alternative frequently used in physics. ■

The distribution given by (8.99) is a solution to the differential equation (8.96), which can be verified using direct substitution. Finding the distributional derivative of the both sides of (8.96), we get

$$\frac{\partial w(x,t)}{\partial x} = -c\delta(x-t), \quad \frac{\partial w(x,t)}{\partial t} = c\delta(x-t), \tag{8.100}$$

and

$$\frac{\partial^2 w(x,t)}{\partial x^2} = -c\delta'(x-t), \quad \frac{\partial^2 w(x,t)}{\partial t^2} = -c\delta'(x-t), \tag{8.101}$$

in which we used the distributional relation

$$\frac{dH(x)}{dx} = \delta(x). \tag{8.102}$$

Substituting the two results of (8.101) into the differential equation (8.96), we confirm that (8.99) is a solution.

[30] A shock wave is a special kind of sonic wave, which is observed when a sound source moves faster than the speed of sound.

Chapter 9

Completion

9.1 Completion of number space

9.1.1 To make it complete artificially

When considering an infinite sequence (of points, numbers, or functions) contained in a space, we would like the limit of the sequence to be contained in the same space. In other words, a space is desired to be **complete.**[1] This is because completeness of a given space guarantees the existence of the limit in the space; as a result of the fact, we are able to establish a self-consistent theory based on the infinite sequence contained in the space.

The advantage of complete spaces will be clarified by studying the properties of an incomplete space. A typical example of incomplete spaces is an open set Ω embedded in $\mathbb{R}^n$. Suppose that a Cauchy sequence in Ω converges to a point just on the boundary of Ω. Due to the openness of Ω, the limit of the Cauchy sequence is not contained in Ω. Hence, we have no way of exploring the limiting behavior of the sequence in a self-consistent manner using only the language of Ω.

To avoid this undesirable situation, let us impose an interesting solution: add forcefully all the limits into the original open set. By forcing the inclusion, we obtain a complete set,[2] denoted $\overline{\Omega}$, which is slightly larger than Ω. **Completion** is the space-control procedure outlined above, which makes an incomplete set be complete by adding deviant limits.

---

[1]A space X is called complete if every **Cauchy sequence** in X has a limit that is also contained in X. Here, a **Cauchy sequence** is a sequence whose elements become arbitrarily close to one another as the sequence progresses. See Appendix B for details.

[2]The renewal set $\overline{\Omega}$ is called the **closure** of Ω; See section 2.2.3.

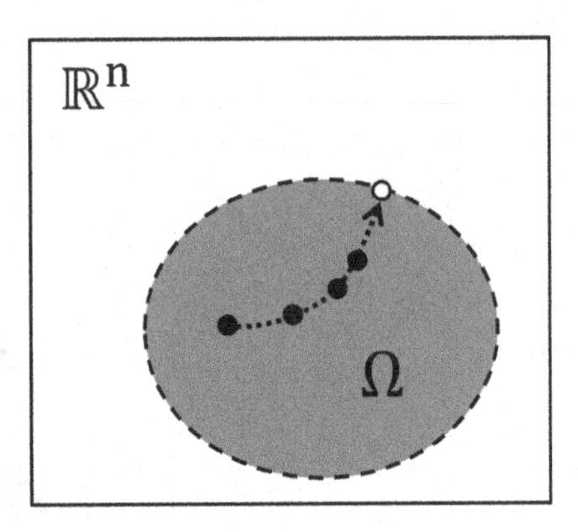

Figure 9.1: Open set Ω embedded in $\mathbb{R}^n$. The infinite point sequence (solid circle) converges to a point (open circle) just on the boundary of Ω.

> **Keypoint:** Completion is a technique of making an incomplete space to be complete.

The story does not end. In the study of functional analysis, most spaces we will examine are abstract spaces (of points or functions), not the three-dimensional Euclidean space that is familiar with us in our daily experience. These abstract spaces do not always have clear, closed boundary curves that separate the insides and outsides of their spaces. Accordingly, even when we say the limit of a Cauchy sequence deviates from the space, we do not mean that the limit protrudes from the closed boundary curve that surrounds the space. Instead, we mean that the distance (topological or normed) inherent in the space does not apply to the limit. Therefore, completion must extend the concept of distance so that it applies to the limit. See section 9.2.1 for a concrete example.

9.1.2 Incomplete space $\mathbb{Q}$

The set of rational numbers, commonly represented by $\mathbb{Q}$, is not a complete set. This implies that the limit of an infinite sequence in $\mathbb{Q}$ may protrude from $\mathbb{Q}$. For instance, π is an irrational number but is the limit of the infinite sequence contained in $\mathbb{Q}$, written by

$$\{3,\ 3.1,\ 3.14,\ 3.141,\ 3.1415,\ 3.14159,\ \ldots\}. \tag{9.1}$$

In order to make $\mathbb{Q}$ complete, therefore, it suffices to add all the non-compliant elements, such as π, to $\mathbb{Q}$ and create a new (somewhat larger) set of numbers. In this manner, we obtain a complete set that we call the set of real numbers, represented by $\mathbb{R}$.

Keypoint: The set $\mathbb{R}$ is a result of completion of $\mathbb{Q}$.

In the following, we have a closer look to the relation between $\mathbb{Q}$ and $\mathbb{R}$ with putting an emphasis on the way how the complete set $\mathbb{R}$ is built from the incomplete set $\mathbb{Q}$.

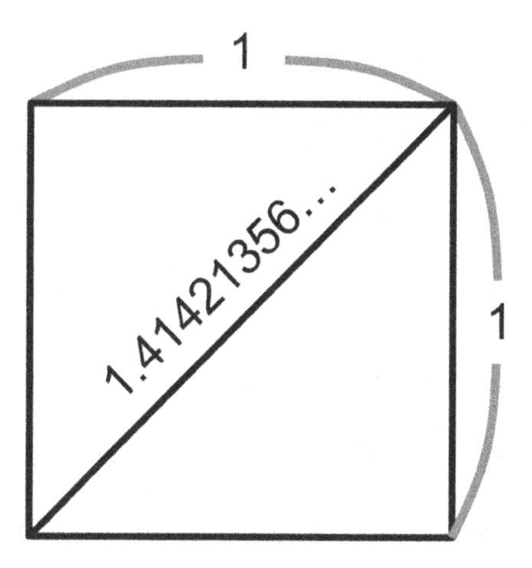

Figure 9.2: A square with sides one unit long. The length of the square's diagonal equals to the positive solution of $x^2 = 2$.

To begin with, we consider a positive solution to the equation $x^2 = 2$. The solution is expressed using the notation $\sqrt{2}$. We know this value is more than 1 and less than 2, and furthermore, it is not a rational number.[3] An alternative view is obtained from elementary geometry. Namely, the positive solution for $x^2 = 2$ is the length of a diagonal of a square with sides one unit long. If we measure the diagonal's length very accurately, we get a very long decimal:

$$1.41421356\cdots \tag{9.2}$$

[3]The fact that $\sqrt{2}$ is not a rational number can be proved by contradiction as follows. First, we assume that $\sqrt{2}$ is a rational number. This is equivalent to saying that $\sqrt{2}$ can be represented as a fraction: $\sqrt{2} = P/Q$, where P and Q are relatively prime natural numbers. By squaring both sides of the equation, we obtain $P^2 = 2Q^2$, which means that P^2 is an even number. If P^2 is even, then P must be also even. If P were an odd number, it could be written as $P = 2M + 1$ for some integer M, and $P^2 = (2M+1)^2 = 2\left(2M^2 + 2M\right) + 1$. This contradicts the fact that P^2 is even. Because P is even, we set $P = 2N$ for some integer N. By squaring both sides, we obtain $P^2 = 4N^2 = 2Q^2$. This gives us $Q^2 = 2N^2$, which means Q^2 is an even number. By a similar argument, if Q^2 is even, then Q must be also even. Therefore, P and Q must both be even and cannot be relatively prime. This contradicts our hypothesis, and $\sqrt{2}$ cannot be rational.

which continues infinitely and has no recurring portion after the decimal point. Hence, the positive solution for $x^2 = 2$ is not a rational number.[4]

We now agree the properties of the non-terminating (infinite) decimal from another perspective. First, we define an infinite sequence of rational numbers $\{a_n\}$ to be

$$\begin{aligned} a_1 &= 1, \\ a_2 &= 1.4, \\ a_3 &= 1.41, \\ a_4 &= 1.414, \\ a_5 &= 1.4142, \\ \vdots & \quad \vdots \end{aligned} \tag{9.3}$$

which monotonically increases toward the infinite decimal in (9.14). From the definition, we have

$$|a_m - a_n| < \frac{1}{10^k} \text{ for any } m, n > k. \tag{9.4}$$

This inequality implies that $|a_m - a_n| \to 0$ as $k \to \infty$. Therefore, the sequence $\{a_n\}$ is a Cauchy sequence contained in the rational number set $\mathbb{Q}$. We should remember that the limit of the sequence $\{a_n\}$ lies outside of $\mathbb{Q}$, and the set $\mathbb{Q}$ is not complete.

Other than the sequence $\{a_n\}$, there are many rational-number Cauchy sequences that converge to the irrational number $1.41421356\cdots$. For instance, it is possible to construct a new sequence $\{b_n\}$ by adding one to the last place of all terms in $\{a_n\}$. The results read as

$$\begin{aligned} b_1 &= a_1 + 1 = 2, \\ b_2 &= a_2 + 0.1 = 1.5, \\ b_3 &= a_3 + 0.01 = 1.42, \\ b_4 &= a_4 + 0.001 = 1.415, \\ b_5 &= a_5 + 0.0001 = 1.4143, \\ \vdots & \quad \vdots \end{aligned} \tag{9.5}$$

Note that $\{b_n\}$ has the same limit as $\{a_n\}$. This is also true for any sequence that is generated by sequentially adding or subtracting, to the last place of the terms in

[4] Any rational number can be represented by a recurring decimal in base 10. For example, the rational number of 3/7 can be written as $3/7 = 0.428571428571428571\cdots$, in which the six digits "428571" repeat after decimal point.

$\{a_n\}$, the terms of a sequence that converges to zero, as in

$$\begin{aligned}
c_1 &= 1+1, \\
c_2 &= 1.4+\frac{1}{2}, \\
c_3 &= 1.41+\frac{1}{2^2}, \\
c_4 &= 1.414+\frac{1}{2^3}, \\
c_5 &= 1.4142+\frac{1}{2^4}, \\
&\vdots \quad \vdots
\end{aligned} \tag{9.6}$$

or

$$\begin{aligned}
d_1 &= 1-1, \\
d_2 &= 1.4-\frac{1}{2}, \\
d_3 &= 1.41-\frac{1}{3}, \\
d_4 &= 1.414-\frac{1}{4}, \\
d_5 &= 1.4142-\frac{1}{5}. \\
&\vdots \quad \vdots
\end{aligned} \tag{9.7}$$

Both of the two sequences are rational Cauchy sequences that converge to $1.41421356\cdots$, similar to $\{a_n\}$.

Among the many choices, we may feel that $\{a_n\}$ is the most natural and standard. The feeling comes from the fact that $\{a_n\}$ is simply derived from the decimal expansion of $\sqrt{2}$ without auxiliary sequential addition or subtraction as performed to other three sequences. However, from a mathematical perspective, all the four sequences, $\{a_n\}$, $\{b_n\}$, $\{c_n\}$ and $\{d_n\}$, should be treated as equivalent to those that converge to the same limit. In fact, our affinity for $\{a_n\}$ is due to our regular use of the decimal system in daily life. But the nature of sequences does not depend on which numbering system we use. In other words, we should not fall into the trap of using a specific numbering system. After discarding this base-10 bias, we see that all Cauchy sequences converging to the same limit are equal; the notion of a better or worse sequence no longer makes sense. Such a unified viewpoint for apparently different Cauchy sequences is a basis for understanding the relationship between $\mathbb{Q}$ and $\mathbb{R}$ demonstrated below.

9.1.3 Creating $\mathbb{R}$ from $\mathbb{Q}$

Assume that we only know about the rational numbers and are completely ignorant of the existence of irrational numbers. Given two classes of rational Cauchy sequences,

$$a_1, a_2, a_3, a_4, \cdots \text{ and } b_1, b_2, b_3, b_4, \cdots, \tag{9.8}$$

can we determine whether these two sequences converge to the same limit? Our question is answered by the following theorem.

Theorem:
Suppose that $\{a_n\}$ and $\{b_n\}$ are Cauchy sequences composed of rational numbers. The two sequences converge to the same limit as $n \to \infty$ if and only if they satisfy the condition

$$|a_n - b_n| \to 0 \ (n \to \infty). \quad \blacksquare \tag{9.9}$$

Proof:
First, we will prove the necessity. If $a_n, b_n \to x$ for $n \to \infty$, we have

$$|a_n - b_n| \leq |a_n - x| + |x - b_n| \to 0 \, (n \to \infty). \tag{9.10}$$

Next, we will consider the sufficiency. Assume that $|a_n - b_n| \to 0$ as $n \to \infty$. Also assume that the two Cauchy sequences $\{a_n\}$ and $\{b_n\}$ converge to x and y, respectively. Then we have

$$|x - y| \leq |x - a_n| + |a_n - b_n| + |b_n - y| \to 0 \, (n \to \infty). \tag{9.11}$$

This means that $x = y$. Therefore, the two sequences converge to the same limit. $\blacksquare$

The most noteworthy fact is that this theorem tells us nothing about the value of the limit. Therefore, it gives no clue as to whether the limit is rational or irrational; the theorem guarantees only the existence of a particular number (an irrational number) to which two or more rational Cauchy sequences converge.

In this context, we again consider the earlier four rational Cauchy sequences, $\{a_n\}$, $\{b_n\}$, $\{c_n\}$ and $\{d_n\}$; See (9.3-9.7). The above theorem states that the four sequences converge to a common number, regardless of whether the number is rational or irrational. Furthermore, there are infinitely many other sequences that converge to the same number (i.e., the positive root of $x^2 = 2$) common to the four sequences mentioned above. In other words, the value of the common limit (that we conventionally write as $\sqrt{2}$) is primarily determined by numerous Cauchy sequences. We thus conclude that the irrational number $\sqrt{2}$ is not primordial but generated by infinitely many sequences of rational numbers having the identical limit. This conclusion holds true for all irrational numbers like $\sqrt{3}$, $\sqrt{5}$, and $\sqrt{2357}$. Every irrational number should not be considered primordial, but as the common limit of many rational Cauchy sequences.

> **Keypoint:** One irrational number is created by infinitely many rational-number sequences.

We understand that irrational numbers did not exist at the beginning. Instead, each irrational number is defined by the common limit of an infinite number of Cauchy sequences in $\mathbb{Q}$. Similarly, the set of all real numbers $\mathbb{R}$ does not exist to begin with. Instead, the set $\mathbb{R}$ is generated by completing the set $\mathbb{Q}$ — by adding the limits of all rational Cauchy sequences to $\mathbb{Q}$. As a consequence, we successfully create a more extensive space (the set $\mathbb{R}$) using only the limited knowledge of a smaller space (the set $\mathbb{Q}$). This is a typical example of completion, the mathematical technique for developing a new and more extensive space from a narrower space.

9.2 Completion of function space

9.2.1 Continuous function spaces are incomplete

In this section, we address the completion procedure for function spaces. As a simple example, we consider an infinite sequence $\{f_n\}$ of continuous functions $f_1(x), f_2(x), f_3(x)\ldots$ that are defined by

$$f_n(x) = \begin{cases} 0, & 0 \le x < \frac{1}{2} - \frac{1}{n+2} \\ (n+2)\left(x - \frac{1}{2} + \frac{1}{n+2}\right), & \frac{1}{2} - \frac{1}{n+2} \le x \le \frac{1}{2} \\ 1, & \frac{1}{2} < x \le 1 \end{cases} \tag{9.12}$$

The sequence $\{f_n\}$ is a Cauchy sequence contained in the function space[5] $C[0,1]$. The neighboring elements f_n and f_{n+1} get infinitely close to each other as n increases. Figure 9.3 demonstrates the approaching behavior of the elements. Note that for an arbitrary n, the elements f_n are continuous over the entire interval $[0,1]$.

The limit of this Cauchy sequence is written as

$$\lim_{n\to\infty} f_n(x) = \begin{cases} 0, & 0 \le x < \frac{1}{2} \\ 1, & \frac{1}{2} \le x \le 1 \end{cases}. \tag{9.13}$$

This limit is a discontinuous function, showing a discrete jump at $x = \frac{1}{2}$. Due to the discontinuity, this limit does not belong to the function space $C[0,1]$. Hence, $C[0,1]$ is an incomplete space. Although the infinite sequence $\{f_n\}$ is completely contained in $C[0,1]$, its limit protrudes from $C[0,1]$.

Keypoint: Continuous function spaces are incomplete.

[5] The symbol $C[0,1]$ represents the functional space that consists of all the continuous functions in the closed interval $[0,1]$.

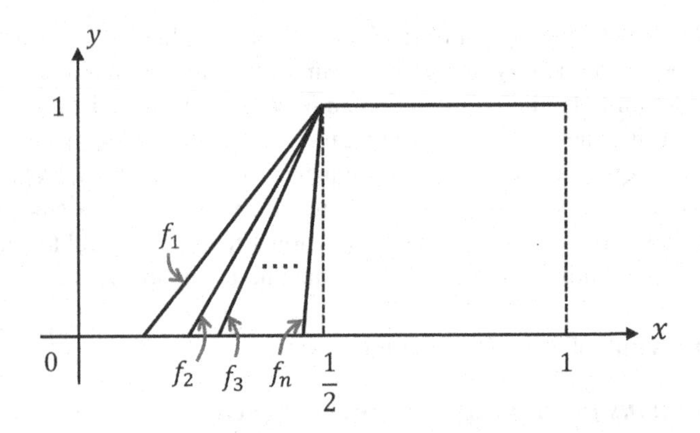

Figure 9.3: Profiles of the sequence $\{f_n(x)\}$ given by (9.12).

The incompleteness of $C[0,1]$ may pose the following question: what kind of function space do we get from the completion of $C[0,1]$? The answer is a **Lebesgue space**, commonly represented by $L^1[0,1]$. The space $L^1[0,1]$ is a function space in which the distance between elements in $L^1[0,1]$ is quantified by the **Lebesgue integral**. Therefore, the presence of a finite number of discontinuous points does not hinder integration. These facts intuitively imply that the completion of $C[0,1]$ is $L^1[0,1]$.

9.2.2 Distance between continuous functions

Let us examine the procedure for generating $L^1[0,1]$ from $C[0,1]$ through completion. In general, when introducing a function space, we need to define the distance between elements. The most important thing to remember is that a wide variety of choices exist for the definition of the distance between elements. Different rules for distance result in different mathematical spaces.

For example, suppose that we introduce a space $C[0,1]$, which consists of all continuous functions defined on the closed interval $[0,1]$. For a pair of elements f and g in the space, it is possible to set the distance $d_M(f,g)$ between them as

$$d_M(f,g) = \max_{0\leq x\leq 1} |f(x) - g(x)|. \tag{9.14}$$

But this is not a unique definition; we can set another rule for the distance between these elements, denoted $d_I(f,g)$, to be

$$d_I(f,g) = \int_0^1 |f(x) - g(x)|\, dx. \tag{9.15}$$

It does not make sense to say that one definition is more correct than the other. Both rules of distance are well defined, and we only need to choose which one is convenient for the problem we are considering.

The nature of a function space depends on our choice of the distance between its elements. Even though two spaces are composed of identical elements, this does not mean they are the same space. The two spaces differ if they are endowed with different rules for determining the distance between elements. This is the reason that different notations are necessary for spaces endowed with different rules for distance. Henceforth, we denote by[6] $C_M[0,1]$ the set of continuous functions defined in $[0,1]$ with distance d_M, while $C_I[0,1]$ with distance d_I.

9.2.3 Completion from C to L^1

Suppose two functions $f(x)$ and $g(x)$ in $C_M[0,1]$ are sufficiently close to each other using the distance d_M. We say that $g(x)$ is located in the neighborhood of $f(x)$. The ε-neighborhood of $f(x)$, written $V_\varepsilon^{[M]}(f)$, is a set of functions neighboring f, using distance d_M, and is expressed as

$$V_\varepsilon^{[M]}(f) = \{g : d_M(f,g) < \varepsilon\}. \tag{9.16}$$

Plainly speaking, $g \in V_\varepsilon^{[M]}(f)$ is equivalent to

$$|f(x) - g(x)| < \varepsilon \tag{9.17}$$

over the interval $0 \le x \le 1$. The upper panel in Fig. 9.4 gives a graphic explanation of $V_\varepsilon^{[M]}(f)$. All elements $g(x)$ should be contained in the region sandwiched between the two curves $y = f(x) \pm \varepsilon$ for any arbitrary x in $[0,1]$. In order for $g(x)$ to belong to $V_\varepsilon^{[M]}(f)$, even a slight portion of the curve $y = g(x)$ cannot deviate from the $\pm\varepsilon$ region. This is the reason a function $h(x)$, whose corresponding curve has a sharp peak that bursts from the $\pm\varepsilon$ region, as is depicted in the bottom panel in Fig. 9.4, is not a member of $V_\varepsilon^{[M]}(f)$.

Consider the case $C_I[0,1]$. When we say $f(x)$ is near $g(x)$ with distance d_I, we mean that $g \in V_\varepsilon^{[I]}(f)$ or, equivalently,

$$\int_0^1 |f(x) - g(x)|\,dx < \varepsilon. \tag{9.18}$$

This inequality implies that the area of the region surrounded by the curves $y = f(x)$, $y = g(x)$, and the vertical lines $x = 0$ and $x = 1$ is smaller than ε. It should be emphasized that in this case, the distance between functions is measured using integration. This rule for distance is completely different from that in the case of $C_M[0,1]$; See (9.14). Hence, the spiky function $h(x)$, which was excluded from $V_\varepsilon^{[M]}(f)$ in $C_M[0,1]$, can be a member of $V_\varepsilon^{[I]}(f)$. Even if the curve $y = h(x)$ has a significantly large peak, we simply need to reduce the peak width of $h(x)$ enough to ensure that the area protruding from $y = f(x)$ is sufficiently small.

---

[6] The subscript M in C_M is the initial "max" used in the definition of the distance d_M.

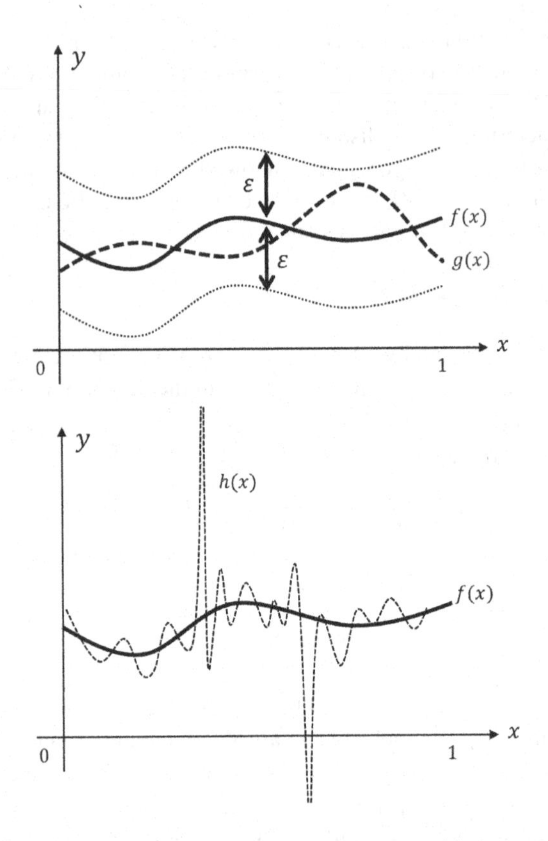

Figure 9.4: Top: In uniform convergence to $f(x)$, any element $g(x)$ should be contained in the region sandwiched between the two curves $y = f(x) \pm \varepsilon$. Bottom: In the sense of L^1, the limit of a spiky function $h(x)$ can converge to $f(x)$.

We continue this examination of the spiky function $h(x)$ in $C_I[0,1]$ by taking the limit at the point where the peak height $h(x)$ diverges and the peak width vanishes. We obtain a discrete function showing divergence at this point. Because $h(x)$ is not a continuous function, it should not be a member of the function space[7] $C_I[0,1]$. Therefore, we complete $C_I[0,1]$ by adding to $C_I[0,1]$ the limiting functions that are excluded from $C_I[0,1]$. This completion generates a new, wider space of functions, called $L^1[0,1]$.

9.2.4 *Strategy for completion*

Thus far we have discussed two specific examples of completion: the completion of $\mathbb{Q}$ that yields $\mathbb{R}$, and the completion of $C_I[0,1]$ that produces $L^1[0,1]$. Many other

[7]This situation is similar to the case of $\mathbb{Q}$, in which the limit of sequences in $\mathbb{Q}$ is not a member of $\mathbb{Q}$.

examples of completion are available, in which we can artificially introduce a new, convenient function space (of points, numbers, or functions) and apply it in order to solve a particular mathematical problem.

In general, the completion procedure is initiated by the following.

1. Begin with an incomplete space endowed with a certain distance between elements.
2. Introduce another definition of distance that is suitable for solving the given problem.

When applying step 1 to a function space, we prefer to begin with a space composed of well-behaved functions, such as smooth, continuous functions. This choice simplifies the mathematical manipulation of the functions, such as differentiation, integration, and integral transformation. It also helps us intuitively understand the calculations. On the other hand, a broad definition of the distance introduced in step 2 is favorable because it will apply to a large class of functions (not exclusively smooth and continuous). Such a broad applicability of distance will relax the restrictions of the types of function classes that can be used. This gives an advantage when developing the theory based on the space and distance defined by the steps 1 and 2.

For example, in the completion of $C_I[0,1]$, the distance d_I was chosen instead of d_M because it is applicable to discontinuous functions. However, a broad definition of the distance also requires us to include a larger number of functions in the space, as was the case with the spiky function $h(x)$. Accepting these functions, which resulted from broadening the concept of distance, as proper members of the space is the mathematical technique we call completion.

9.3 Sobolev space

9.3.1 What is Sobolev space?

The completion of an incomplete (normed) space allows us to introduce a new, artificially designed function space that is suitable for our specific purposes. A pivotal example of this is the **Sobolev space**, which is specifically powerful in solving partial differential equations. The main feature of a Sobolev space is found in its definition of the norm, which is a combination of the L^p **norms** of a function in the space and its derivatives up to a given order. In addition, the derivative used in the definition of the norm is not the usual one but that used for **distributions**, called a **weak derivative** (see section 8.5.1). Sobolev spaces are important because solutions to partial differential equations are found naturally in Sobolev spaces rather than in spaces of continuous functions with classically understood derivatives.[8]

Let us attempt to generate a Sobolev space using the two-step procedure mentioned in section 9.2.4. In the first step, we choose an initial function space that is

[8]For a detailed mathematical background of Sobolev spaces, refer to the book by Haim Brezis, "Functional Analysis, Sobolev Spaces and Partial Differential Equations," Springer, 2011.

as simple as possible. Here, we choose the space[9] $C^k(\Omega)$, which consists of smooth, continuous functions of degree k defined on an open set $\Omega \subset \mathbb{R}$. In the second step, we define a norm. Given a function f in the space $C^k(\Omega)$, we define the norm of f to be

$$\begin{aligned}\|f\|_{H^k} &= \left(\int_\Omega |f(x)|^2 dx + \int_\Omega |Df(x)|^2 dx + \int_\Omega \left|D^2 f(x)\right|^2 dx + \cdots + \int_\Omega \left|D^k f(x)\right|^2 dx\right)^{1/2} \\ &= \left(\sum_{s=0}^{k} \int_\Omega |D^s f(x)|^2 dx\right)^{1/2}. \qquad (9.19)\end{aligned}$$

Here, the symbol D indicates the **distributional derivative**. It is known that the space $C^k(\Omega)$ with norm $\|f\|_{H^k}$ is not complete. If we observe the converging behavior of an infinite sequence in $C^k(\Omega)$, the limit may extend beyond the scope of $\|f\|_{H^k}$. In order to complete the space, the missing points that cannot be caught by the norm $\|f\|_{H^k}$ should be added to $C^k(\Omega)$. As a result, $C^k(\Omega)$ is enlarged to a Lebesgue space. After performing some manipulations, the following complete function space shall be obtained.

Definition (Sobolev space):
Consider a function $f \in L^p(\Omega)$ defined on an open set $\Omega \subset \mathbb{R}$. Assume that f is differentiable in the sense of distribution up to the kth degree and that the derivatives, designated by $D^s f$ $(0 \le s \le k)$, also belong to $L^p(\Omega)$. Then a **Sobolev space** $W_p^k(\Omega)$ is defined by the set of functions:

$$W_p^k(\Omega) = \{f \in L^p(\Omega);\ D^s f \in L^p(\Omega), 0 \le s \le k\}, \qquad (9.20)$$

with norm

$$\|f\|_{H^k} = \left(\sum_{s=0}^{k} \int_\Omega |D^s f(x)|^2 dx\right)^{1/2}. \quad \blacksquare \qquad (9.21)$$

Keypoint: Sobolev spaces are special kinds of L^p spaces; the norm is defined through distributional derivatives.

9.3.2 *Remarks on generation of Sobolev space*

We saw that the Sobolev spaces $W_p^k(\Omega)$ are processed from the well-behaved space $C^k(\Omega)$ by introducing the norm $\|f\|_{H^k}$, which involves the weak derivatives of f. Because the limits of infinite sequences in $C^k(\Omega)$ may deviate from the cover of $\|f\|_{H^k}$, we forcefully attach these protruding limits to $C^k(\Omega)$, and the playground

[9] The symbol C^k represents a space of continuous functions that are differentiable up to degree k. A simple example of an element contained in this space is $|x|^{k+1}$ for an even k, which does not have a $(k+1)$st derivative at $x = 0$.

enlarges from $C^k(\Omega)$ to $L^p(\Omega)$. In particular, when $k = 0$, the Sobolev space is equivalent to a Lebesgue space as represented by

$$W_p^0(\Omega) \equiv L^p(\Omega). \tag{9.22}$$

Similarly, when $p = 0$, the Sobolev space is a Hilbert space[10], which we typically write as

$$H^k(\Omega) \equiv W_0^k(\Omega). \tag{9.23}$$

We should remark that the distributional derivative was involved in the definition of Sobolev space. In general, the completion procedure may cause denaturation of mathematical manipulations in the original function. When generating Sobolev spaces, for example, classic differentiation denatures to distributional differentiation. In the realm of functional analysis, such an alteration to the nature of the mathematical manipulations is regarded as a constructive extension, which means it has a positive effect. The original space X is contained in the completed space $\tilde{X}$. Therefore, the manipulations used in X remain active and applicable to a part of the elements in $\tilde{X}$.

9.3.3 *Application to Laplace operator manipulation*

What is the merit in defining Sobolev spaces? Why should we introduce the space in such an artificial manner using the concept of weak derivatives and the complex definition of the norm in (9.19)? The primary answer lies in their roles as domains for differential operators.

In many fields of physics, such as electromagnetics, fluid mechanics, and quantum mechanics, the **Laplace operator**

$$\Delta = \sum_{i=1}^{N} \frac{\partial^2}{\partial x_i^2} \tag{9.24}$$

is heavily used to describe physical phenomena. However, this operator is only applicable to second-order differentiable functions. The Laplace operator cannot be applied to Dirac's delta functions or Heaviside's step functions, both of which are familiar in various theories of physics. This restriction substantially hinders the development of theoretical descriptions.

A breakthrough idea for overcoming this difficulty is to introduce the norm $\|f\|_{H^k}$ in (9.19), which involves the concept of weak derivatives. As a consequence, we create a complete function space, in which the Laplace operator can be applied to any element. This is the reason Sobolev spaces play a vital role in physics and engineering as the domains of Laplace operators.

---

[10]The inner product of f and g in the Hilbert space $H^k(\Omega)$, denoted by $(f,g)_{H^k}$, is defined to be

$$(f,g)_{H^k} = \sum_{s=0}^{k} \int_\Omega D^s f(x) \overline{D^s g(x)} dx,$$

where $\overline{D^s g(x)}$ is the complex conjugate of $D^s g(x)$.

Keypoint: In Sobolev spaces, Laplace operators can apply to any functions, even though they are discontinuous or divergent.

9.3.4 Application in solving differential equations

Another practical significance of Sobolev spaces is their usability as tools for solving a certain class of partial differential equations. For example, suppose we have the **Poisson equation** with respect to $u(x)$ given by

$$\begin{cases} -\Delta u(x) = f(x), \ (x \in \Omega) \\ u(x) = 0, \ (x \in \partial\Omega) \end{cases}. \tag{9.25}$$

Here, $f(x)$ is a given function, Ω is an open set contained in $\mathbb{R}^n$, and $\partial\Omega$ represents the boundary of Ω. The Poisson equation frequently appears in electromagnetics, in which the unknown function $u(x)$ corresponds to the electric potential produced by a charge that is spatially distributed over Ω with density $f(x)$. The second line in (9.25) gives the boundary condition, indicating that the outer edge is connected to the ground and ensuring that the potential along the edge is zero.

In a common-sense view of physics, one experimental setup results in one particular electric potential field. Put others, when we determine the way of earth grounding and electric charge quantity in the apparatus, we will obtain one unique potential field profile, with no multiple realizations of the field. Hence, we may think that the same is true in mathematics. The Poisson equation (9.25) seems to provide a unique solution $u(x)$ unless $f(x)$ is an ill-defined, pathological function. However, this conjecture is not true in general. In fact, the existence of a solution to a Poisson differential equation, as well as the uniqueness of this solution, is not at all trivial. This has long been considered an important problem in the theory of partial differential equations.

Historically speaking, the existence of the Poisson equation has been proved in part by many approaches. One famous strategy, based on the variation principle, states that the solution of (9.25) minimizes the functional $I(u)$ defined by

$$I(u) = \frac{1}{2}\int_\Omega |\nabla u(x)|^2 dx - \int_\Omega f(x)u(x)dx. \tag{9.26}$$

Nevertheless, determining whether the function $u(x)$ that minimizes $I(u)$ is differentiable to the second order is not trivial. Currently, this ambiguity is resolved with the use of weak derivatives, through which we attempt to find solutions to (9.25) using the concept of distributions. Specifically, we address this problem with a step-by-step approach.

1. Let $u(x)$ be an element of a Sobolev space and prove the existence of the optimal $u(x)$ that minimizes $I(u)$ in the Sobolev space.
2. Confirm that $u(x)$ is a weak solution to the Poisson equation.

3. Prove that $u(x)$ is second-order differentiable in the usual sense.

Sobolev spaces let us prove the existence and uniqueness of a Poisson equation's solution. This resolution has been a giant step for subsequent development in modern theory of partial differential equations[11].

Keypoint: Sobolev space-based arguments allow us to prove the existence and uniqueness of a Poisson equation's solution.

[11] Applications of versatile function spaces, including Sobolev spaces, for solving partial differential equations can be found in the book by Lawrence C. Evans, "Partial Differential Equations" (Graduate Studies in Mathematics Vol.19), 2nd edition, American Mathematical Society, 2010.

Chapter 10

Operator

10.1 Classification of operators

10.1.1 Basic premise

An **operator** is a mathematical device that determines the way in which elements in a set are mapped to elements in another set. The simplest example of a mapping is the function f that associates each number y in $\mathbb{R}$ or $\mathbb{C}$ with a number x in $\mathbb{R}$ or $\mathbb{C}$, respectively. This mapping is commonly designated using $y = f(x)$ or $f : x \to y$. In the study of functional analysis, more general types of mappings must be considered. These include mappings between points, numbers, functions, and other abstract mathematical entities. Despite multiplicity of entities we will consider, we would like to continue our discussion using a single perspective. In order to do this, we employ **operator theory**, which is a major branch of functional analysis.

In practice, the mappings and spaces involved must meet certain requirements. For example, the mapping should be **linear** and should relate elements in two **vector spaces**.[1] Furthermore, operators that act on *infinite-dimensional* vector spaces are critically important[2] because they have versatile applications in physics and engineering. Therefore, the primary objective of this chapter is to survey the workings of linear operators that act on infinite-dimensional vector spaces.

---

[1] A **vector space** is a set of element to which basic mathematical operations, such as addition and scalar multiplication, can be defined. A precise definition of vector space is provided in section 3.1.2.

[2] In addition to infinite-dimensional spaces, finite-dimensional spaces are also covered in operator theory. However, given a finite-dimensional space, elementary linear algebra sufficiently describes the linear mappings between vector spaces, in which operators are represented using finite-dimensional matrices.

Keypoint: Although more general types of operators and spaces are addressed in operator theory, we restrict our study to *linear operators* in *vector spaces*.

10.1.2 Norms and completeness

In order to develop a meaningful discussion of infinite-dimensional spaces, the following two concepts must be examined: the **norm** of an element and the **completeness** of a space. A vector space that has both a norm and completeness is called a **Banach space**, which was introduced in section 3.3.4. Moreover, if the inner product property is applied to a Banach space, then it becomes a **Hilbert space**. We will primarily focus on linear operators in infinite-dimensional Banach (and Hilbert) spaces. However, before continuing, we should explain the reason we want norms and completeness to be defined in our vector spaces.

First, we must remember that the vector space axiom provides no criteria for measuring the closeness between two elements in a vector space (see section 3.1.2). Therefore, we have no means of examining the convergence of infinite sequences of elements in a given space. This problem is resolved by inducing a **topology** to the space, which transforms the vector space into a **topological vector space**. Various methods can induce a topology, and different topologies result in different topological vector spaces. A simple topological vector space is called a **normed space**, which is a vector space endowed with a **norm**. The norm generalizes the length of a vector in the classical sense (i.e., the length of a geometric arrow) and is conventionally written $||x||$ for vector x, in which the value of $||x||$ is non-negative. Using the norm, the distance between two elements, x and y, is represented by $||x-y||$. Similarly, we can quantify the distance between two images, $T(x)$ and $T(y)$, in which operator T acts on x and y using $||T(x)-T(y)||$. This notion of distance allows us to measure the closeness between elements and examine the convergence of sequences in the space.

Because inducing the norm in a space is insufficient, we must also address the concept of **completeness**.

The need of complete spaces becomes clear when, for instance, solving differential equations. Given a differential equation, determining the solution in a single stroke can be difficult. Instead, an initial solution is often approximated and then improved upon as repeated approximations of the solution increase its accuracy. The ultimate solution, obtained by repeating the approximations infinitely many times, is the solution to a differential equation.[3]

---

[3]This kind of successive approximation is also used in the theory of quantum mechanics and is known as **perturbation theory**. In quantum mechanics, the state of a quantum particle is described using a linear combination of the solutions to a specific differential equation. However, determining the exact solution to the differential equation is practically impossible. Perturbation theory allows us to find an approximated solution using successive approximations, which increases accuracy and can also truncate the higher order terms in the original differential equation.

Because it is a limit, we should determine whether the ultimate solution is contained in the set of functions under consideration (i.e., it is a **function space**). If a solution exists in our function space, we can create a self-consistent discussion concerning the properties of the differential equation using the norm of the space and the class of functions contained in the space. For example, polynomial expansions of a solution, as well as the distance between two different solutions, are explicitly written in the language of the specific function space. However, this is not always true because, in infinite-dimensional spaces, the limit of a sum or sequence of elements is frequently not contained in the space. In this case, we can no longer examine the mathematical properties of the solution using the rules of the space, nor can we apply the same methods to solving the differential equation. However, this inconvenience is eliminated if the function space is complete because completeness guarantees that the ultimate solution exists within the space. This can be said for many kinds of mathematical problems, in addition to differential equations. On the other hand, when a space is not complete, our discussion of convergence can lose proper meaning.

Keypoint: Our study is further restricted to linear operators in Banach (and Hilbert) spaces.

10.1.3 Continuous operators

In the previous two subsections, we restricted our consideration to linear operators in infinite-dimensional Banach spaces. But even under the restriction, various types of operators are still included in the class of linear operators. So another question remains; Which linear operators do we consider? In order to answer this, we should revisit the reason Banach spaces were chosen as our primary algebraic structure. Because every Banach space is endowed with a norm, this allows us to measure the distance between elements in an intuitive manner. Therefore, it is natural to focus on linear operators that are "**continuous**" (or equivalently "**bounded**")[4] in the sense of our norm.

Therefore, we must understand what it means for a linear operator to be continuous. As we will see later, a **continuous linear operator** maps two distinct elements that are significantly close to each other onto two elements that are also substantially close to each other. Continuous linear operators tend to be more tractable than other classes of linear operators, because their properties are analogous to some degree with those of finite-dimensional matrices in elementary linear algebra. Moreover, continuous linear operators are typically the first operators studied in operator theory.

[4]If we limit our study to linear operators, continuity of the operator is equivalent to boundedness (see section 10.2.5 for details).

10.1.4 Completely continuous operators

Continuous linear operators form the most fundamental class of operators in Banach spaces. The mathematical consequences obtained from applying operations using continuous linear operators on elements in *infinite*-dimensional Banach spaces are similar in many aspects to those obtained from applying operations that use matrices on vectors in *finite*-dimensional vector spaces. However, it should be noted that the two mathematical entities (i.e., a continuous linear operator and a finite-dimensional matrix) are not entirely identical.

For example, the concept of an eigenvalue (and a corresponding eigenvector) does not logically apply to function spaces because the convergence property of point sequences differs for finite-dimensional and infinite-dimensional spaces. In finite-dimensional topological vector spaces, a specific convergent subsequence can always be extracted from an infinite and bounded sequence embedded in the space. Furthermore, the limit of the subsequence is always contained in the same space.[5] These convergence properties on subsequences may or may not be true for infinite dimensions. If we consider an infinite and bounded point sequence in an infinite-dimensional topological vector space, a means of extracting any subsequence that converges to a point in the space may not exist. These contrasting behaviors indicate that the nature of the space drastically alters depending on whether the dimension of the space is finite or infinite.

Therefore, introducing an eigenvalue-like concept cannot be done by imposing a certain condition onto function spaces. Because of the essential differences between finite and infinite spaces, these counter-productive controls would not provide the corresponding concepts of eigenvalue and eigenvector in function spaces. Thus, in order to achieve a possible compromise, we can impose a specific strong condition on the continuous linear operators, rather than on the function space. Later, we will see that when this strong condition is imposed on continuous linear operators, we can create a new class of operators, called **completely continuous linear operators**. Employing this new class of operators allows us to formulate analogous eigenvalues and eigenvectors that work properly in function spaces. It should be noted that only the continuity of linear operators is insufficient when used to describe eigenvalue-like problems in infinite dimensions. Therefore, we need to develop the notion of *complete* continuous linear operators in order to proceed, although the resulting operators are subject to stricter conditions than continuous (not completely continuous) linear operators.

10.1.5 Non-continuous operators

As we previously mentioned, continuous (or completely continuous) operators are the most favorable and tractable for our use, because their properties are relatively

[5] A topological space that has these two convergence properties on subsequences is referred to as a **compact space**. In other words, if a topological space X is compact, then any infinite sequence of points contained in X has an **accumulation point** in X.

intuitive. Thus, it is natural that we would prefer employing only continuous operators when considering function spaces. However, continuous operators are not as versatile as we would expect. In particular, the assumptions contained in the definition of operator continuity are too strong for many practical situations. This is especially true when using operators to solve differential equation or when formulating quantum mechanics. In these subjects, continuous operators do not generally apply.

The two major classes of non-continuous operators of significant practical importance are: (i) differential operators that act on $L^2(\mathbb{R})$ spaces and (ii) simple multiplication operators that map function $f \in L^2(\mathbb{R})$ onto xf with $x \in \mathbb{R}$. The fundamental importance of these operators is understandable. If we cannot use differential operators on $L^2(\mathbb{R})$, we cannot examine the existence of any differential equation with respect to function $f \in L^2(\mathbb{R})$ or the accuracy of its approximate solution. In addition, multiplication and differential operators are the most fundamental operators used in the theory of quantum mechanics. Multiplication operators are used to describe the position of a quantum particle, and differential operators indicate the particle's momentum. These operators (the position and momentum operators) are the most essential operators in quantum mechanics but cannot be captured using the most basic operators in functional analysis (continuous operators).

Keypoint: *Non-continuous* operators have practical importance.

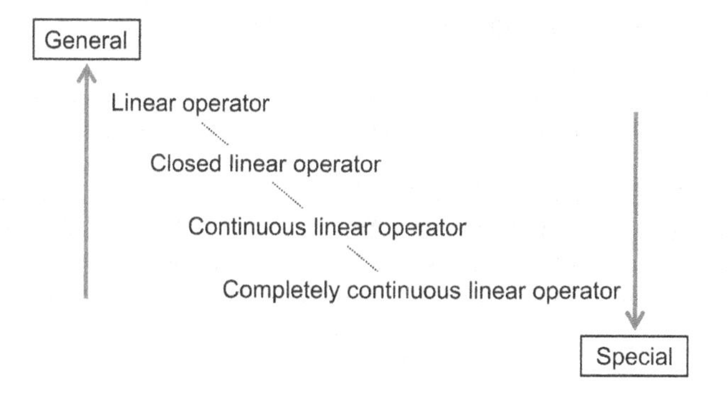

Figure 10.1: Hierarchy of operators used in functional analysis.

10.1.6 Closed operators

Because continuous operators are frequently ineffective when describing mathematical procedures or physical processes, we must look beyond continuous operators, which are easily understandable, to non-continuous operators, which tend to exceed our intuitive understanding. However, dismissing the concept of continuity in linear operators would be wasteful. Therefore, let us impose a moderately weakened condition of continuity onto linear operators, which creates a new class of operators. These resulting linear operators are known as **closed operators**. In the practical application of functional analysis, closed operators play an active role in physical and engineering problems to which continuous operators do not apply. Unfortunately, the analogy to elementary linear algebra does not hold for closed operators because of the loss of continuity. This produces many technical problems in the mathematical manipulation of closed operators and the convergence test for related materials. The complicated mathematical background and methods used for addressing this issue are topics for advanced function analysis.

Although the study of closed operators is important, it is beyond the scope of a functional analysis introduction. In the rest of this chapter, description of closed operators (and other non-continuous linear operators) is kept to a minimum, and we focus on the formulation of continuous linear operators instead.

10.2 Essence of operator theory

10.2.1 Definition of linear operator

We develop the essential theory of linear operators in Hilbert spaces with infinite dimension. To begin with, the precise definition of linear operators and related materials are demonstrated.

Definition (Linear operator):
Let X and Y be topological vector spaces with infinite dimension, and D be a subspace of X. Assume that an operator T associates every $u \in D$ uniquely to $T(u) \in Y$. If T satisfies

$$T(\alpha u+\beta v)=\alpha T(u)+\beta T(v), \quad (u,v\in D, \quad \alpha,\beta\in\mathbb{C}) \tag{10.1}$$

T is called a **linear operator** acting on D. ■

Under the condition mentioned above, we call the subspace D by the **domain** of T. The notation $D(T)$ is also used if we would like to represent explicitly that D is the domain of the operator of T. In accord with, the subspace

$$R(T)=\{T(u):\ u\in D(T)\} \tag{10.2}$$

is called the **range** of the operator T. Particularly when $D(T)$ and $R(T)$ are both subspaces of X, we say that T is an operator *in* X.

Observe that the linearity of operators, defined by (10.1), has nothing to do with neither notions of norm, limit, nor convergence of a certain sequence. Hence, linear operators can apply to general topological vector spaces, not limited to Banach spaces. The generality disappears when we define the continuity and boundedness of linear operators. To introduce the continuity and boundedness of some mathematical entities, we need the concepts of distance between elements and convergence of infinite sequences of elements. Therefore, the spaces on which continuous or bounded linear operators are defined must be endowed with norm and completeness.

10.2.2 Operator's domain

In infinite dimension, the nature of a linear operator is dependent significantly on the choice of its domain. For instance, when the differential operator $T = -id/dx$ is given, we can choose its domain $D(T)$ simply as[6]

$$D(T) = C^\infty(\mathbb{R}) \tag{10.3}$$

or a more limited space as

$$D(T) = \{f \in C^\infty(\mathbb{R})\,; f(x) \text{ is a periodic function with the peroid of } 2\pi\}. \tag{10.4}$$

The former choice means that we allow the operator T to act on any functions that are infinitely differentiable. In the latter choice, the operation of T is restricted only to periodic functions. In either case, the mathematical representation of the operator $T = -id/dx$ is invariant. Nevertheless, the difference in the choice of its domain $D(T)$ may cause a drastic alteration in the consequences obtained by operating T on the functions in the domain.[7] Therefore, we should keep in mind that, whenever we use an operator acting on an infinite-dimensional space, we must specify clearly what subspace is chosen as the domain of the operator.

The above-mentioned caution on the choice of domains is relevant even when we examine the equality of two linear operators. Let S and T be both linear operators that map X to Y, in which both X and Y are topological vector spaces with infinite dimension. If we want to tell whether the two operators S and T are equal or not, we need to examine whether the domains $D(S)$ and $D(T)$ are completely identical to each other. It does not suffice to check whether S and T are in the same mathematical representation.

Particularly when the relationships

$$D(S) \subset D(T) \quad \text{and} \quad S(u) = T(u) \tag{10.5}$$

hold simultaneously for every $u \in D(S)$, we say that the operator T is an **extension** of the operator S, and used the notation of $S \subset T$. It is obvious

$$S = T \iff S \subset T \text{ and } T \subset S. \tag{10.6}$$

---

[6]Here, $C^\infty(\mathbb{R})$ is the set of all functions that are infinitely differentiable and defined on $\mathbb{R}$.

[7]This situation is in completely contrast with the case of finite-dimensional spaces, in which the nature of a linear operator do not depends on the choice of its domain.

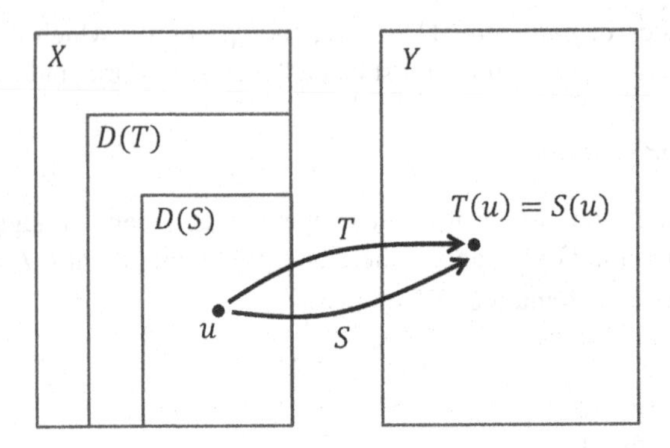

Figure 10.2: Diagram of the relations: $D(S) \subset D(T)$ and $S(u) = T(u)$ presented in (10.5).

10.2.3 *Definition of continuous linear operator*

Among many kinds of linear operators, continuous linear operators are the first to be learned in functional analysis. A preliminary notion is given below.

Definition (Locally continuous linear operator):
Let T be a linear operator that relates two Banach spaces, X and Y. We say that T is **locally continuous** at a point $u_0 \in D(T)$, provided the following conditions are satisfied.

1. There exists an infinite sequence of points $\{u_n\}$ that are contained in $D(T)$.
2. The sequence $\{u_n\}$ converges to u_0 in the norm of X, namely,

$$\|u_n - u_0\|_X \to 0. \quad (n \to \infty) \tag{10.7}$$

3. The image of the sequence $\{T(u_n)\}$ converges to $T(u_0)$ in the norm of Y, namely,

$$\|T(u_n) - T(u_0)\|_Y \to 0. \quad (n \to \infty) \quad \blacksquare \tag{10.8}$$

Notice that in the definition above, we assumed both X and Y to be Banach spaces. This was because the norms and convergences in X and Y were required as written in (10.7) and (10.8).

Plainly speaking, local continuity of a linear operator T means that the convergence at a specific point is preserved before and after the mapping. Actually the definition above requires that, if an infinite sequence of points $\{u_n\}$ converges to u_0, then the sequence should be mapped onto another infinite sequence $\{T(u_n)\}$ that converges to $T(u_0)$.

Interestingly, this convergence-preserving property of a locally continuous linear operator is quite analogous to that of a locally continuous function $f(x)$ at a specific point. Local continuity of a curve $y = f(x)$ at $x = x_0$ means that, if a sequence of points along the x axis converges to x_0, the sequence should be mapped to another point sequence lying on the y axis that converges to $y_0 = f(x_0)$. If the local continuity of the curve $y = f(x)$ holds at arbitrary points on the x axis, we say that f is a continuous function. In this context, it is natural to give the following definition on the continuous linear operator.

Definition (Continuous linear operator):
If T is locally continuous at everywhere in $D(T)$, we say that T is a **continuous linear operator**. ■

Two examples below show continuous and non-continuous linear operators.

Example 1:
Let K be a continuous function on $[0,1] \times [0,1]$, and T be an operator in the function space of $L^2([0,1])$ defined by

$$T:\ u(x) \to \int_0^1 K(x,y)u(x)dx,\ \text{where}\ u \in L^2([0,1]). \tag{10.9}$$

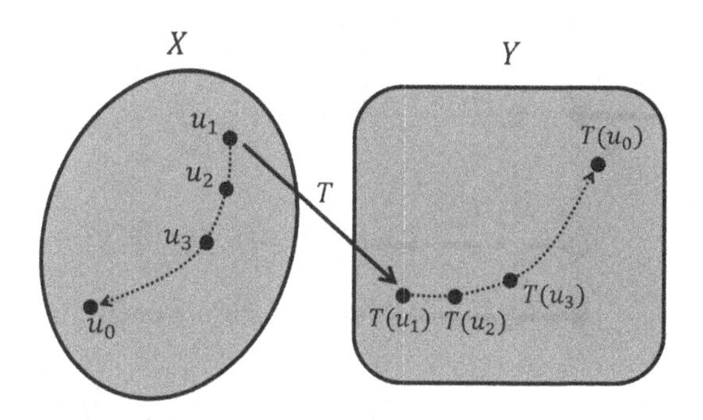

Figure 10.3: A continuous operator $T : X \to Y$ maps a converging sequence $\{u_n\}$ in X onto a converging sequence $\{T(u_n)\}$ in Y.

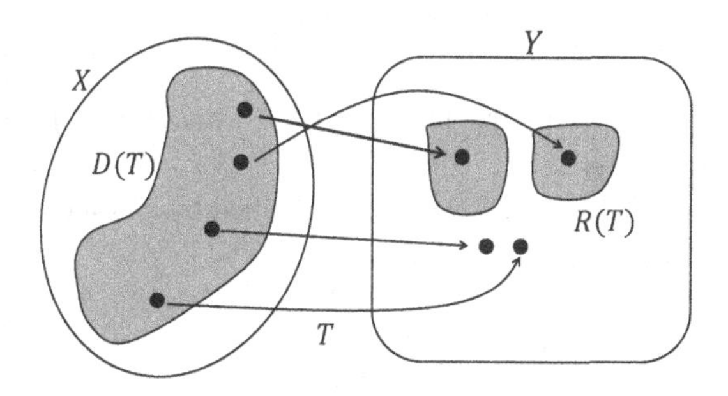

Figure 10.4: When an operator $T : X \to Y$ is not continuous, two neighboring points in $D(T)$ may not be mapped onto two neighboring points in $R(T)$. Namely, a non-continuous operator does not preserve the convergence property.

In this case, T is a continuous linear operator that maps $L^2([0,1])$ to $L^2([0,1])$. ■

Example 2:
Consider a subspace of $C^\infty\left((0,1)\right)$ defined by[8]

$$C_0^\infty\left((0,1)\right) = \left\{f \in C^\infty\left((0,1)\right) : \text{supp}(f) \text{ is bounded}\right\}. \tag{10.10}$$

Here, supp(f) indicates the **support** of f, which is the closure of a set Ω of points at which $f(x) \neq 0$;

$$\text{supp}(f) = \overline{\Omega}, \ \Omega \equiv \left\{x : f(x) \neq 0\right\}. \tag{10.11}$$

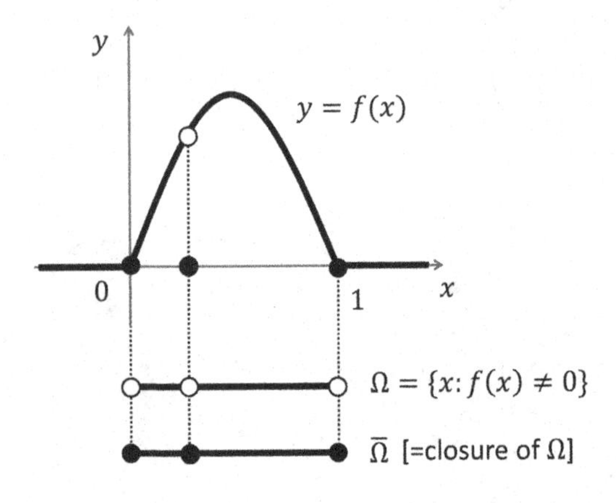

Figure 10.5: Support of f is defined by the closure of a set $\Omega = \{x : f(x) \neq 0\}$.

Let T be a differential operator with the domain $D(T) = C_0^\infty\left((0,1)\right)$, as written by

$$T(u) = \frac{du}{dx}, \quad u \in C_0^\infty\left((0,1)\right). \tag{10.12}$$

In this case, T is a linear operator that maps $L^2\left((0,1)\right)$ to $L^2\left((0,1)\right)$. However, it is not continuous in $D(T)$. ■

The conclusion of Example 2 above will be clarified by considering a sequence of functions $\{u_n\}$ defined by

$$u_n(x) = \frac{x^n}{\sqrt{n}}. \quad (n = 1, 2, \cdots) \tag{10.13}$$

[8]Here, $C^n\left((0,1)\right)$ is the set of n-times differentiable functions defined on the open interval $(0,1)$.

Observe that $u_n \in C_0^\infty\left((0,1)\right)$ for any n. The L^2-norm of u_n reads

$$\|u_n\| = \sqrt{\frac{1}{n}\int_0^1 x^{2n}dx} = \frac{1}{\sqrt{n(2n+1)}} \to 0. \quad (n \to \infty) \tag{10.14}$$

Namely, the sequence $\{u_n\}$ converges to the zero element represented by $u(x) \equiv 0$ at $x \in (0,1)$, and the limit is contained in $D(T)$. If the differential operator T were continuous at zero, therefore, the image of the sequence $\{T(u_n)\}$ should have converged to the zero element in $R(T)$ [i.e., in $L^2\left((0,1)\right)$], too. But this is not the case because the L^2-norm of $T(u_n)$ reads

$$\|T(u_n)\| = \sqrt{n\int_0^1 x^{2n-2}dx} = \sqrt{\frac{n}{2n-1}} \to \frac{1}{\sqrt{2}}. \quad (n \to \infty), \tag{10.15}$$

Hence, T is not continuous locally at the zero element. Accordingly, it is not a continuous operator.

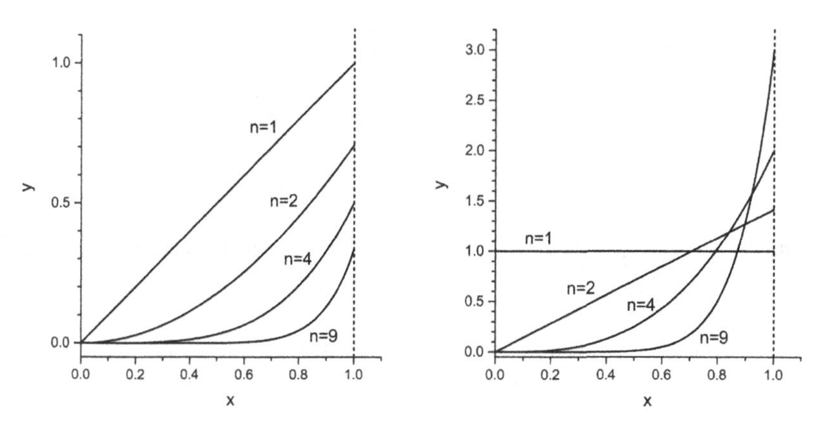

Figure 10.6: Left: Profiles of the function $u_n(x) = x^n/\sqrt{n}$ within $0 < x < 1$. Right: Profiles of the first derivative of $u_n(x)$, given by $du_n/dx = \sqrt{n}x^{n-1}$, within $0 < x < 1$.

10.2.4 Definition of bounded operator

The concept of continuity in a linear operator is closely related to the notion of boundedness defined below.

Definition (Bounded operator):
An operator T (not necessarily continuous) that maps one Banach space X onto the other Banach space Y is **bounded** if, for any $u \in D(T)$, there exists $M > 0$ such that

$$\|T(u)\|_Y \leq M\|u\|_X. \quad \blacksquare \tag{10.16}$$

The boundedness of an operator T indicates that the image of u, denoted by $T(u)$, is not jumping out to place extremely far away from the zero element in $R(T)$, but remains staying nearby zero to a certain degree. It should be emphasized that the boundedness property is independent of the continuity of an operator. In fact, a bounded operator may or may not be linear. Similarly, a linear operator may or may not be bounded.

Keypoint: Boundedness has no relation to linearity. Indeed, ...
Linear operators *may or may not* be bounded, and
Bounded operators *may or may not* be linear.

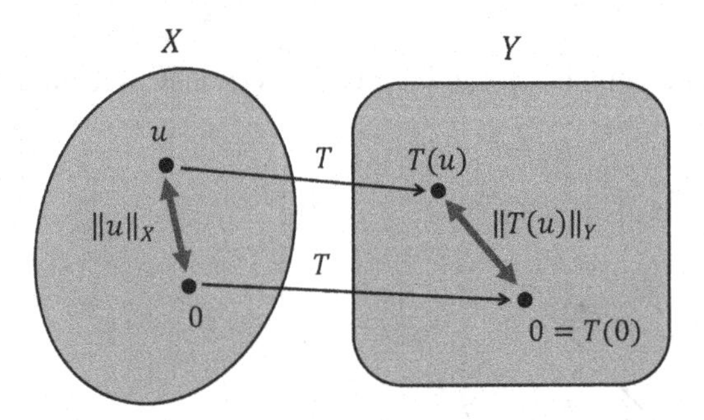

Figure 10.7: A bounded operator $T : X \to Y$ maps a pair of two neighboring points in X onto a pair of neighboring points in Y.

Using the concept of **operator norm**, which is defined below, the definition of boundedness of the operator T is reduced to a more simplified expression.

Definition (Operator norm):
Let $T : X \to Y$ be a continuous linear operator. Its **operator norm** $\|T\|$ is defined by

$$\|T\| = \frac{\|T(u)\|_Y}{\|u\|_X}. \quad \blacksquare \tag{10.17}$$

Theorem: $\|T\| < \infty$ if and only if T is bounded on the domain $D(T)$. ■

10.2.5 Equivalence between continuity and boundedness

The linearity of an operator T provides a conceptual bridge between the continuity and boundedness. As a matter of fact, the two concepts become equivalent if and only if T is linear. Refer to the formal statement below.

> **Theorem:**
> Continuous linear operators are bounded.
> Bounded linear operators are continuous. ■

Proof:
Consider the linear operator $T : X \to Y$. If T is continuous, it is locally continuous at 0. What we would like to show is that, if T is locally continuous at 0, then T is bounded. This will complete the proof of the first half of the theorem above.

Let us perform the proof by contradiction. To begin with, we assume that T is not bounded. In the instance, there exists $u \in D(T)$ for which (10.16) fails no matter how large M is chosen. Accordingly, we can find a point sequence u_n that satisfies

$$\|T(u_n)\|_Y > n\|u_n\|_X. \quad (n = 1, 2, ...) \tag{10.18}$$

The inequality indicates that[9] $u_n \neq 0$ for all n. Since $\|u_n\|_X \neq 0$ for all n, it is possible to divide u_n by $\sqrt{n}\|u_n\|_X$ to obtain

$$v_n = \frac{u_n}{\sqrt{n}\|u_n\|_X}, \tag{10.19}$$

which implies

$$\|v_n\|_X = \frac{1}{\sqrt{n}} \to 0. \quad (n \to \infty) \tag{10.20}$$

On the other hand, the linearity of T implies that

$$\|T(v_n)\|_Y = \frac{\|T(u_n)\|_Y}{\sqrt{n}\|u_n\|_X} > \sqrt{n} \to \infty. \quad (n \to \infty) \tag{10.21}$$

The two results, (10.20) and (10.21), suggest that T should not be continuous at 0. This conclusion is contradictory to our assumption that T is locally continuous at 0. In summary, the first half of the theorem has been proved.

Next we address the second half of the theorem: "*If T is bounded, it is continuous.*" The boundedness of T implies that for arbitrary $u \in D(T)$ and $u_n \in D(T)$, we have

$$\|T(u_n - u)\|_Y \leq M\|u_n - u\|_X. \tag{10.22}$$

Furthermore, the linearity of T implies

$$\|T(u_n) - T(u)\|_Y = \|T(u_n - u)\|_Y. \tag{10.23}$$

---

[9] The validity of $u_n \neq 0$ for all n under the condition of (10.18) is confirmed by the following argument. Assume that $u_n = 0$ at a certain n. Since T is linear, we have $T(0) = T(0+0) = T(0) + T(0)$, which implies $T(0) = 0$. Namely, it is concluded that $T(u_n) = 0$ at the specific n, which conflicts (10.18).

Combine the two results to obtain

$$\|T(u_n) - T(u)\|_Y \leq M\|u_n - u\|_X. \tag{10.24}$$

Taking the limit of $\|u_n - u\|_X \to 0$ $(n \to \infty)$ in the right side of (10.24), we observe that for any u,

$$\|T(u_n) - T(u)\|_Y \to 0. \quad (n \to \infty) \tag{10.25}$$

This indicates that T is continuous in $D(T)$. ■

Let us summarize the present argument. In the first half of the proof above, we have shown that

A linear operator T is continuous
$\Rightarrow$ T is locally continuous at 0 $\Rightarrow$ T is bounded.

After that, we have demonstrated in the second half of the proof that

A linear operator T is bounded $\Rightarrow$ T is continuous.

Therefore, what we have confirmed is the fact that a linear operator T is continuous if and only if it is locally continuous at 0. Namely, we have proved that

T is locally continuous at 0 $\Leftrightarrow$ T is entirely continuous.

Emphasis should be placed on that the local continuity of a linear operator at a certain point is equivalent to the continuity of the operator over the domain. As far as we are considering linear operators, the local continuity assures the global continuity. This equivalent between local and global continuity is the salient feature of linear operators. In fact, neither continuous functions $y = f(x)$ nor general non-linear operators is endowed with the local-global continuity equivalence.

Keypoint: For linear operators, local and global continuities are equivalent.

10.2.6 Artificial downscale of operator domain

As earlier explained, linear operators are preferred to be bounded (or equivalently continuous) from a viewpoint of easiness to deal in. So what would be a good idea if we encounter a linear operator that is not bounded? One possible idea is that we artificially scale down the domain of the non-bounded operator in order for it to be bounded.

For instance, let T be a non-bounded linear operator, whose domain is entirely over an infinite-dimensional vector space X. To produce a bounded operator, we artificially pick up a finite number of elements $x_1, ..., x_n$ from X. These elements span a finite-dimensional vector space X_n. It is known that every linear operator on finite-dimensional space is bounded. Therefore, we can make T be bounded by downscaling the domain of T from X to X_n. A practical example of this downscaling procedure is given below.

Example:
Let $H^1(0,1)$ be a **Sobolev space**, and T be the differential operator in the sense of distribution that acts on $f \in H^1(0,1)$ such as

$$Tf = \frac{df(x)}{dx}. \tag{10.26}$$

Note that T is not bounded in the domain of $D(T) = H^1(0,1)$. To produce a bounded operator, we prepare a m-dimensional subspace of $D(T)$, defined by

$$X_m = \left\{ f : f(x) = \sum_{k=1}^{m} a^k \sqrt{2} \sin(k\pi x), \; a^k \in \mathbb{C} \right\}. \tag{10.27}$$

Next we downscale the domain from $D(T)$ to X_m, and denote the resulting operator by T_m. The operator norm of T_m reads

$$||T_m|| = m\pi. \tag{10.28}$$

Since $||T_m||$ does not diverge, T_m is bounded in X_m. Finally we obtain a bounded linear operator T_m. ■

The domain-downscaling technique for creating a bounded operator is quite useful when we want to approximate a non-bounded operator that is hard to deal with as it is. The efficiency of the technique is highlighted in numerical computation of functions and operators. In numerical computations, it is almost impossible to manipulate mathematical entities with infinite degree of freedom. We thus often compromise by approximate calculation of up to sufficiently high dimension. A typical example of the approximation is observed in Fourier analysis by numerical calculations. In principle, Fourier series expansion is to calculate infinite series of trigonometric functions (or imaginary exponentials). In numerical computation, we abandon the infinitely many summations; instead, we truncate the summation at a sufficiently high order and secure an approximated series expansion. In this process, the original infinite-dimensional space spanned by trigonometric functions is downscaled into a finite-dimensional space by truncation.[10]

10.3 Preparation toward eigenvalue-like problem

10.3.1 Spectrum

Solving the **eigenvalue problem** is the central topic in elementary linear algebra that deals in finite-dimensional vector spaces. Surprisingly, the problem becomes

[10]Of course, this truncation technique is beneficial only when discarded terms give minor contributions to the problem we are considering.

meaningless in infinite dimension. Given a linear operator in a function space, it is generally impossible to establish the notion of eigenvalues and eigenvectors of the operator. The breakdown of the notion in function spaces is a crucially important manifestation that straightforward analogy from the elementary linear algebra is not valid in infinite dimension.

An apparently possible idea of the concept expansion may be based on the proportional relation of

$$Tx = \alpha x \quad (x \neq 0). \tag{10.29}$$

Namely, we would assume that for a given linear operator T, the vector x and the scholar α that satisfy (10.29) are regarded as the eigenvector and eigenvalue of T. But this idea turns out to be totally invalid, because infinite-dimensional vector spaces generally do not possess such element x that satisfies (10.29). Hence, we can no longer rely on the proportional relation in order to induce an eigenvalue-like concept in infinite dimension.

So, what would we do? An alternative idea is to look for the scholar $\alpha \in \mathbb{C}$ that makes the operator $T - \alpha I$ to be **invertible.**[11] Here, we assume that T is a continuous linear operator that transforms a vector in a Banach space to another vector in the identical Banach space. The operator I is the identity operator in the same space. If we collect all possible values of α that satisfy the condition above, then the set of those α is known to build a closed set contained in $\mathbb{C}$. This set of specially chosen α is called the **spectrum** of the continuous linear operator T. Particularly when T is *completely* continuous, the spectrum behaves in a similar manner to the set of eigenvalues of finite-dimensional matrices.[12] This is why we are saying that completely continuous operators are the most essential for beginners studying functional analysis.

10.3.2 Self-adjoint operator

In order to accomplish the spectrum-eigenvalue correspondence, we need to focus only operators in Hilbert spaces, instead of Banach spaces. Furthermore, only a special class of linear operators, called **self-adjoint operators**, is allowed to be dealt with as outlined below.

Suppose that a linear operator T transforms a function $f \in L^2([0,1])$ to xf. Actually this operator is known to be a self-adjoint operator. In addition, it can be proved that[13] the spectrum of T is the closed interval $[0,1]$. Importantly, when α

---

[11] By saying that a linear operator T in a Banach space is **invertible**, we mean that there exists an **inverse operator** of T in the same Banach space, denoted by S, which satisfies the relation: $TS = ST = I$.

[12] In contrast, operators that are continuous but not completely continuous give rise to conceptual discrepancy between spectrum (for infinite dimension) and eigenvalue (for finite dimension).

[13] Observe that if α locates out of the interval $[0,1]$, $x - \alpha$ cannot be zero since $0 \leq x \leq 1$. In this case, there exists an **inverse operator** of $T - \alpha I$, which maps $f(x)$ to $f(x)/(x-\alpha)$. This fact indicates that $[0,1]$ is the spectrum of T.

lies within the interval, we can realize the situation that[14] the image of the operator, Tf, behaves in an almost same manner to the operator αf. Note that this situation is analogous to the product $\boldsymbol{A}\boldsymbol{v}$ of a matrix $\boldsymbol{A}$ with a vector $\boldsymbol{v}$ in finite dimension. Through diagonalization technique, the product $\boldsymbol{A}\boldsymbol{v}$ is decomposed to a set of the products $a_i\boldsymbol{e}_i$, where a_i and $\boldsymbol{e}_i$ are the ith eigenvalue of $\boldsymbol{A}$ and its corresponding eigenvector, respectively. The analogy holds only when the operator is self-adjoint.

Keypoint: Spectrum-eigenvalue correspondence holds only if the operator is *linear, completely continuous, self-adjoint* and it acts on *Hilbert spaces.* (= Extremely strict conditions must be imposed to both operators and spaces!)

10.3.3 Definition of self-adjoint operator

Let us introduce an **adjoint operator** of a given linear operator T. The adjoint operator is, so to say, an operator version of an adjoint matrix with finite dimension. In elementary linear algebra, the **adjoint matrix** of an m-by-n complex matrix $\boldsymbol{A}$ is the n-by-m matrix $\boldsymbol{A}^*$ obtained from $\boldsymbol{A}$ by taking the transpose and then taking the complex conjugate of each entry. For example, if

$$\boldsymbol{A} = \begin{bmatrix} 1 & -2-i \\ 1+i & i \end{bmatrix}, \tag{10.30}$$

then we have

$$\boldsymbol{A}^* = \begin{bmatrix} 1 & 1-i \\ -2+i & -i \end{bmatrix}. \tag{10.31}$$

So what kind of operators does adjoint matrices correspond to? Since no matrix-entry is assigned to operators, transposition-based definition is of no use for our aim. Instead, we pay attention to the following property of general adjoint matrices.

$$(\boldsymbol{A}\boldsymbol{u}, \boldsymbol{v}) = (\boldsymbol{u}, \boldsymbol{A}^*\boldsymbol{v}) \tag{10.32}$$

We make a concept-expansion on the basis of (10.32), which is the relation between two inner products. The need of inner product for the definition of adjoint matrices implies the need of inner product for defining their operator versions, too. This is why we consider operators in Hilbert spaces as candidates for operators having adjoint properties. Below is an exact definition of adjoint operators.

Definition (Adjoint operator):
Let T be a linear operator in a Hilbert space. The **adjoint operator** T^* of T is defined by

$$(T(u), v) = (u, T^*(v)) \tag{10.33}$$

[14]In order to realize the situation, we need to choose such $f \in L^2([0,1])$ that yields $f(x) = 0$ almost everywhere and $f(x) \neq 0$ only at the vicinity of $x = \alpha$. Under this condition, Tf and αf are effectively the same operators, similarly to the relation between $\boldsymbol{A}\boldsymbol{e}_i$ and $a_i\boldsymbol{e}_i$ commented above.

The domain of T^* is defined by[15]

$$D(T^*) = \left\{ v \in X : \begin{array}{c} \text{For any } u \in D(T), \text{ there exists } w \in X \\ \text{such that } (T(u), v) = (u, w) \end{array} \right\}. \quad \blacksquare \tag{10.34}$$

The essential properties of the adjoint operator T^* of T are listed below.

1. T^* is a linear operator acting on the domain $D(T^*) \subset X$.
2. $(T(u), v) = (u, T^*(v))$ for $u \in D(T)$, $v \in D(T^*)$
3. $\|T^*\| = \|T\|$.

We are in a position to introduce the notion of **self-adjoint operators**. In elementary linear algebra, self-adjoint matrices (or called Hermite matrices) are defined by those which satisfy the relation

$$\boldsymbol{A} = \boldsymbol{A}^*. \tag{10.35}$$

Straightforward extension of this definition allows us to introduce their "operator" version as follows.

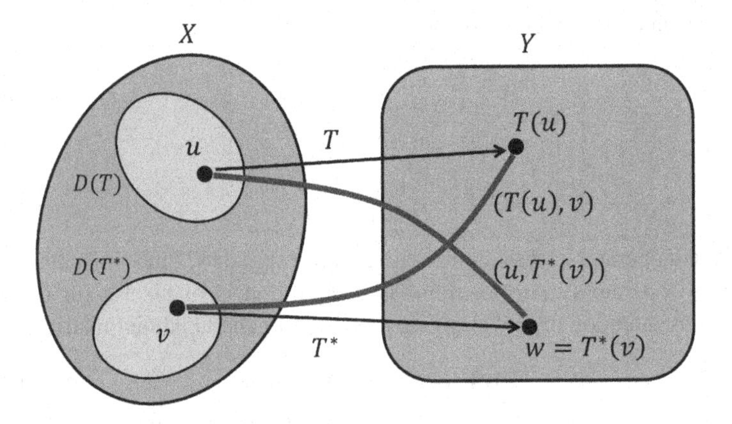

Figure 10.8: Diagram of the adjoint operator T^* of T.

Definition (Self-adjoint operator):
Suppose that T is a linear operator in a Hilbert space X and the domain $D(T)$ satisfies $\overline{D(T)} = X$. If $T = T^*$, we call T as a **self-adjoint operator**. $\blacksquare$

In the above description, the overbar in $\overline{D(T)}$ indicates to take a **closure** of $D(T)$.

[15] If the required w exists, it is determined uniquely by v. In fact, if $w_1, w_2 \in X$ and $(T(u), v) = (u, w_1) = (u, w_2)$ for any $u \in D(T)$, then it is concluded that $w_1 = w_2$.

10.3.4 Compact or non-compact?

Profound disagreements in the nature of eigenvalue problems between *finite-* and *infinite*-dimensional spaces originates from the difference in the **compactness** of the spaces. The notion of compactness is a generalization of the closedness in Euclidean space, i.e., the property that all points in the spaces lie within a certain finite distance of each other.

The precise definition of compactness follows from the argument below. Suppose that an infinite sequence of points is contained in a space. We want to extract an infinite "sub"sequence from the original sequence, under the condition that the subsequence must converge to a point in the space considered. If we can find such a subsequence no matter how the original infinite sequence is set, the space is called to be compact.

As an example, consider the n-dimensional number space $\mathbb{C}^n$ and a unit sphere S_c^n embedded in $\mathbb{C}^n$:

$$S_c^n = \left\{(z_1, z_2, \cdots, z_n) : |z_1|^2 + |z_2|^2 + \cdots + |z_n|^2 = 1\right\}. \tag{10.36}$$

The set of points S_c^n is known to be compact, because a converging subsequence can be picked up from arbitrary infinite point sequences on S_c^n. In contrast, a unit sphere S with infinite dimension, represented by

$$S = \{x : \|x\| = 1\} \tag{10.37}$$

is not compact. Roughly speaking, infinite-dimensional spaces are not closed but open to infinitely many directions. Therefore, subsequences that were extracted from an infinite sequence may not be convergent.

Keypoint: Sequences on S_c^n are compact. But those on S are non-compact.

In-depth discussion on the subject above-mentioned will lead us to the theory of eigenvalue analyses for infinite dimension. The theory obtained is called the **spectral theory**, which is one of the central branch in the study of functional analysis.

Why S is not compact, although S_c^n is compact? The non-compactness of S is accounted for by the divergence property of orthonormal bases on S. Let $\{e_1, e_2, \cdots, e_n, \cdots\}$ be an orthonormal basis of the space S. It means that every element located on S is written by a linear combination of the unit vectors $e_1, e_2, \cdots, e_n, \cdots$. It is to be noted that any pair of e_m and e_n with $m \neq n$ yields

$$\|e_m - e_n\|^2 = \|e_m\|^2 - (e_m, e_n) - (e_n, e_m) + \|e_n\|^2 = 1 - 0 - 0 + 1 = 2, \tag{10.38}$$

which implies that

$$\|e_m - e_n\| = \sqrt{2} \text{ for arbitrary pairs of } m, n. \tag{10.39}$$

Accordingly, we can find infinitely many sequences of points on S in which all pairs of points in the sequences are apart by $\sqrt{2}$ each other. In other words, there exists on S an infinite number of points, all of which are equi-separated. No matter what subsequence are picked up from the original sequence, therefore, the subsequence should no longer be convergent. This is the reason why S is not compact.

10.3.5 Loss of eigenvectors in infinite dimension

What impact does the non-compactness of S have on our argument? Surprisingly, the non-compactness of S causes the inability of defining operator's eigenvalues and eigenvectors in S.

Let us take a closer look on the issue. Suppose that H is a linear, bounded, and self-adjoint operator in an infinite-dimensional Hilbert space. We further assume that H satisfies[16]

$$(Hx,x) \geq 0. \tag{10.40}$$

It is seemingly natural that the relation

$$Hx_0 = \lambda x_0, \quad x_0 \neq 0 \tag{10.41}$$

should provide a way of determining the eigenvalue of H and the corresponding eigenvector, denoted by λ and x_0, respectively. If this idea were true, λ must be a non-negative real number.[17] Yet our intent fails as explained below.

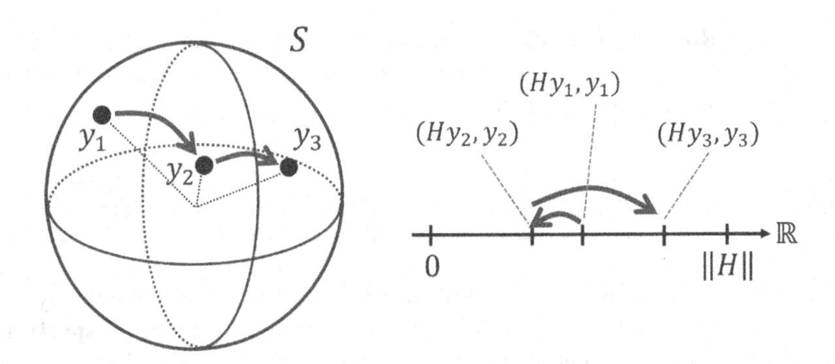

Figure 10.9: Migration of y_i on S results in a variation in the value of (Hy_i,y_i) along the real axis.

[16]The assumption (10.40) is imposed only for simplicity of the following argument. If we want to consider the case of $(Hx,x) \leq 0$, we have only to reverse the sign of H in our argument based on the assumption of $(Hx,x) \geq 0$.

[17]This readily follows from $(Hx_0,x_0) = \lambda(x_0,x_0)$ and the assumption of (10.40).

In the meantime, we try to find the maximum eigenvalue of H in accord with the superficial definition of (10.41). For all x, we have

$$0 \leq (Hx,x) \leq \|Hx\| \cdot \|x\| \leq \|H\| \cdot \|x\|^2. \tag{10.42}$$

Divide each term by $\|x\|^2$ to obtain

$$0 \leq \left(H\frac{x}{\|x\|}, \frac{x}{\|x\|} \right) \leq \|H\|, \tag{10.43}$$

and set $y = x/\|x\|$. Realize that every point y lies on S, since $\|y\| = 1$. In addition, (Hy,y) is a continuous function with respect to y, which satisfies the magnitude relation of

$$0 \leq (Hy,y) \leq \|H\|. \tag{10.44}$$

These arguments seemingly imply that, we are allowed to define the maximum eigenvalue of H (denoted by $\lambda_{\max}$ if it exists) by

$$\lambda_{\max} \equiv \max_{y \in S}(Hy,y). \tag{10.45}$$

Moreover, it seems correct that the vector y that makes the value (Hy,y) be maximum is regarded as an eigenvector of H. But this prospect fails because the non-compact nature of S prevents us from finding y that makes (Hy,y) be maximum.

Why are we not able to find the required y_0? To comprehend the reason, we imagine that the point y is traveling over S. The wandering of y on S results in a variation in (Hy,y), while the value of (Hy,y) is restricted within the closed interval $\left[0, \|H\|\right]$ on the real number axis. Consider the supremum (i.e., the least upper limit) of (Hy,y), and denote it by

$$\overline{\lambda} \equiv \sup_{y \in S}(Hy,y). \tag{10.46}$$

Since $\overline{\lambda} \leq \|H\|$, $\overline{\lambda}$ is a candidate for the maximum eigenvalue of H. Moreover, since $\overline{\lambda}$ is the supremum of (Hy,y), there exists a point sequence $\{y_1, y_2, \cdots, y_n, \cdots\}$ on S that satisfies

$$\overline{\lambda} - (Hy_n, y_n) < \frac{1}{n} \quad \text{for any } n = 1, 2, \cdots. \tag{10.47}$$

Under the condition, (Hy_n, y_n) approaches $\overline{\lambda}$ unlimitedly as n increases. Still yet, the sequence $\{y_n\}$ does not converge to any point on S due to the loss of compactness of S. This fact means that there exists no vector y_n which is in one-to-one correspondence to $\overline{\lambda}$. In conclusion, it is impossible to find on S such y that makes (Hy,y) be maximum.

Keypoint: There exists no eigenvector for operators that act on infinite-dimensional spaces.

10.4 Eigenvalue of "completely continuous" operator

10.4.1 Condition imposed on continuous operator

Our observation in the previous subsection holds true for general infinite dimensional spaces. We cannot any more rely on the superficial definition (10.41) for seeking eigenvalues and eigenvectors for operators.

Then what should we do? As far as S is non-compact, no contrivance for improving the convergence property of $\{y_n\}$ would be successful. So we change the way of thinking. Instead of $\{y_n\}$, we rely on the point sequence $\{Hy_n\}$ for a breakthrough. We explore the condition under which a convergent subsequence $\{Hy_{n_i}\}$ $(n_i = 1, 2, \cdots)$ can be extracted from $\{Hy_n\}$ $(n = 1, 2, \cdots)$ that lies in S. If we successfully obtain a convergent subsequence $\{Hy_{n_i}\}$, then it provides a way of constructing eigenvalues and eigenvectors of the operator H.

In order to complete the attempt suggested above, we prepare the following condition that shall be imposed to continuous linear operators H.

"Condition ♣" for a continuous operator H:
Let $\{x_n\}$ be an infinite point sequence on S.
If we pick up an appropriate subsequence $\{x_{n_i}\}$ from $\{x_n\}$, then the subsequence is mapped by an operator H to $\{Hx_{n_i}\}$ that is convergent on S.

In addition to the above-mentioned condition imposed on H (hereafter we call it by "the condition ♣"), we hypothesize ongoingly that $(Hx, x) \geq 0$ for any x [see

(10.40)]. It is known that the self-adjoint property of H implies[18]

$$\|H\| = \sup_{y \in S}(Hy, y). \tag{10.49}$$

As a consequence, we finally obtain the theorem below.

Theorem:
If a self-adjoint operator H in S is linear, bounded, and satisfies the condition ♣, the operator possess an eigenvalue λ and eigenvector $x_0(\neq 0)$ such as

$$Hx_0 = \lambda x_0. \quad \blacksquare \tag{10.50}$$

From the theorem, it is understood that the condition ♣ is very encouraging to construct the notion of eigenvalues and eigenvectors in infinite dimension. In this context, it is fair to say that the condition ♣ provides a critical clue to resolve the discrepancy between finite and infinite dimensions.

Proof:
The theorem is trivial if H is the zero operator that takes all vectors into 0. Hereafter we assume $H \neq 0$. Setting $\lambda = \|H\|$, we see from (10.49) that

$$\lambda = \sup_{y \in S}(Hy, y) > 0. \tag{10.51}$$

---

[18] Below is a proof of $\|H\| = \sup_{y \in S} |(Hy, y)|$, instead of (10.49). The proof is completed by confirming the two inequalities:

$$\|H\| \geq \sup_{y \in S} |(Hy, y)| \text{ and } \|H\| \leq \sup_{y \in S} |(Hy, y)|. \tag{10.48}$$

The former inequality is easily shown using **Schwarz's inequality**: $|(Hy, y)| \leq \|Hy\| \cdot \|y\| \leq \|H\| \cdot \|y\|^2 = \|H\|$ for $\|y\| = 1$. To prove the latter inequality, we set $y = x/\|x\|$ for all x except at $x = 0$. Substituting it to $\gamma \equiv |(Hy, y)|$, we have $-\gamma\|x\|^2 \leq (Hx, x) \leq \gamma\|x\|^2$. This result is also true for $x = 0$. As a result, we obtain for any y_1, y_2 satisfying $\|y_1\| = \|y_2\| = 1$,

$$\Big(H(y_1 + y_2), y_1 + y_2\Big) = (Hy_1, y_1) + (Hy_2, y_2) + 2\mathrm{Re}\Big[(Hy_1, y_2)\Big] \leq \gamma\|y_1 + y_2\|^2.$$

A parallel argument to $\Big(H(y_1 - y_2), y_1 - y_2\Big)$ yields

$$\Big(H(y_1 - y_2), y_1 - y_2\Big) = (Hy_1, y_1) + (Hy_2, y_2) - 2\mathrm{Re}\Big[(Hy_1, y_2)\Big] \geq -\gamma\|y_1 - y_2\|^2.$$

Hence we have

$$4\mathrm{Re}\Big[(Hy_1, y_2)\Big] \leq \gamma\left(\|y_1 + y_2\|^2 + \|y_1 - y_2\|^2\right) = 2\gamma\left(\|y_1\|^2 + \|y_2\|^2\right) = 4\gamma.$$

Set $y_1 = y$ and $y_2 = Hy/\|Hy\|$, and substitute them to obtain $\|Hy\| \leq \gamma$. Since $\|H\| = \|Hy\| \leq \gamma$ under the present condition, we finally obtain $\|H\| \leq |(Hy, y)|$. As a result, both the two inequalities in (10.48) have been verified.

The definition of the supremum implies that there exists an appropriate point sequence $\{y_n\}$ that yields

$$\lambda = \lim_{n\to\infty} (Hy_n, y_n). \tag{10.52}$$

As well, the condition ♣ implies that there exists a subsequence $\{y_{n_i}\}$ such that $\{Hy_{n_i}\}$ converges to a point on S. For simplicity, we replace the labels n_i by n and write the limit of $\{Hy_n\}$ as

$$\lim_{n\to\infty} Hy_n = x_0 \in S. \tag{10.53}$$

Using the self-adjoint property of H, we have

$$\begin{aligned} \|Hy_n - \lambda y_n\|^2 &= \|Hy_n\|^2 + \lambda^2 \|y_n\|^2 - \lambda\,(Hy_n, y_n) - \lambda\,(y_n, Hy_n) \\ &= \|Hy_n\|^2 + \lambda^2 \|y_n\|^2 - 2\lambda\,(Hy_n, y_n) \\ &\to \|x_0\|^2 + \lambda^2 - 2\lambda^2 = \|x_0\|^2 - \lambda^2. \quad (n\to\infty) \end{aligned} \tag{10.54}$$

In the derivation, we used $\|y_n\|^2 = 1$. Since $\|Hy_n - \lambda y_n\|^2$ in the leftmost term of (10.54) is non-negative,

$$\|x_0\|^2 \geq \lambda^2 > 0. \tag{10.55}$$

Hence $x_0 \neq 0$. Next, we substitute the relation $\|Hy_n\|^2 \leq \|H\|^2 \cdot \|y_n\|^2 = \lambda^2$ into the second line of the right side in (10.54) to obtain

$$\lim_{n\to\infty} \|Hy_n - \lambda y_n\|^2 \leq \lambda^2 + \lambda^2 - 2\lambda^2 = 0. \tag{10.56}$$

It indicates that $\lim_{n\to\infty} \|Hy_n - \lambda y_n\| = 0$, namely, the sequence $\{Hy_n - \lambda y_n\}$ converges to zero. Since H is continuous, the sequence $H\{Hy_n - \lambda y_n\}$ also converges to zero, as written by

$$\lim_{n\to\infty} H\,(Hy_n - \lambda y_n) = 0. \ (n\to\infty) \tag{10.57}$$

Combine the result with (10.53) to obtain

$$\lim_{n\to\infty} H\,(Hy_n) = \lim_{n\to\infty} \lambda H y_n = \lambda x_0. \tag{10.58}$$

It is thus concluded that

$$Hx_0 = \lambda x_0. \tag{10.59}$$

According to (10.55), λ is considered as an eigenvalue of H. ■

10.4.2 Definition of completely continuous operator

We have just learned that the condition ♣ strongly supports our formulation of the eigenvalue problems for continuous and self-adjoint operators. It should be emphasized that the condition ♣ can be imposed on general continuous operators A, not limited to continuous and self-adjoint operators. The condition guarantees the convergence of Ax_{n_i}, wherein $\{x_{n_i}\}$ is an infinite subsequence extracted from $\{x_n\}$ that satisfies $\|x_n\| = 1$ for any n. If a continuous operator is endowed with this property, then it is referred to as **complete continuous**, as stated below.

Definition (Complete continuous operator):
Continuous operators that satisfy the condition ♣ are called **completely continuous operators.** ■

Using the terminology, we can say that given an operator has eigenvalues and eigenvectors if it is both completely continuous and self-adjoint.

Interestingly, the identity operator I is one of the most typical "counter"-example of completely continuous operators. To see it, let $\{e_1, e_2, \cdots, e_n, \cdots\}$ be an orthonormal basis of a Hilbert space. Since $Ie_n = e_n$ for any n and $\|e_n - e_n\| \equiv 2$ for any pairs of n and $m(\neq n)$, it is impossible to extract a convergent subsequence from $\{Ie_n\}$.[19] The situation drastically alters when we consider complete continuous operators. Indeed, provided A is a completely continuous operator, we become able to extract a convergent subsequence from a point sequence $\{Ae_n\}$ lying on S. In infinite-dimensional spaces, therefore, completely continuous operators file a position that is totally opposite with that of the identity operator.

Keypoint: Eigenvalues/Eigenvectors of an operator exist only when the operator is both complete continuous and self-adjoint.

10.4.3 Decomposition of Hilbert space

Let λ_n be the n-th eigenvalue of the operator H that is completely continuous and self-adjoint. If we arrange all the eigenvalues in order of decreasing their absolute values, we have the magnitude relation

$$\|H\| = |\lambda_1| \geq |\lambda_2| \geq \cdots \geq |\lambda_n| \geq \cdots \to 0. \tag{10.60}$$

Here, 0 may or may not be included in the set of eigenvalues; if it is included, we denote it by $\lambda_\infty = 0$. As a preparatory remark, we remind that two eigenspaces $E(\lambda_m)$ and $E(\lambda_n)$ each of which belongs to different eigenvalues, λ_n and $\lambda_{m(\neq n)}$, are orthogonal to each other.

The following theorem will be used in our subsequent argument.

Theorem (Decomposition for eigenspaces):
Let H be a completely continuous and self-adjoint operator that acts on Hilbert spaces. Every Hilbert space Ψ can be decomposed to a set of eigenspaces of H such as[20]

$$\Psi = E(\lambda_1) \oplus E(\lambda_2) \oplus \cdots \oplus E(\lambda_n) \oplus \cdots. \quad ■ \tag{10.61}$$

[19] The absence of convergent subsequence in $\{Ie_n\}$ is just a restatement that the infinite-dimensional unit sphere S is not compact.

[20] If 0 is contained in the eigenvalues of H, (10.61) is replaced by $\Psi = E(\lambda_1) \oplus \cdots \oplus E(\lambda_\infty)$, in which $E(\lambda_\infty)$ appears at the leftmost term.

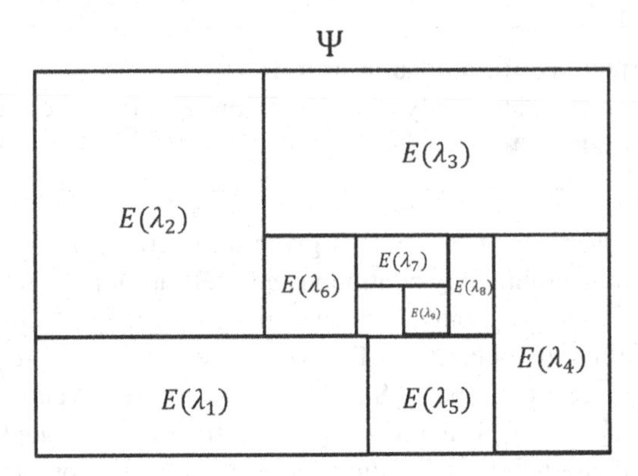

Figure 10.10: Decomposition of a Hilbert space Ψ into a set of eigenspaces $\{E(\lambda_n)\}$.

What is implied by the theorem is explained below. Suppose that each eigenspace of H, denoted by $E(\lambda_n)$, has the dimension of $k_n = \dim E(\lambda_n)$. For each eigenspace, we can set an orthonormal basis that consists of k_n unit vectors. The basis is written by[21]

$$\left\{ e_1^{(n)}, e_2^{(n)}, \ldots, e_{k_n}^{(n)} \right\}. \tag{10.62}$$

Remind that the basis of (10.62) spans the single eigenspace $E(\lambda_n)$. In turn, the set of eigenspaces $\{E(\lambda_n)\}$ spans the whole Hilbert space, as stated in the theorem above. Therefore, what the theorem implies that every element x in Ψ is represented uniquely by a linear combination of infinitely many unit vectors as

$$x = \sum_{n=1}^{\infty} \sum_{i=1}^{k_n} \alpha_i^{(n)} e_i^{(n)}, \quad \sum_{n=1}^{\infty} \sum_{i=1}^{k_n} \left| \alpha_i^{(n)} \right|^2 < +\infty. \tag{10.63}$$

10.4.4 Eigenvalue of completely continuous operator

We are ready to accomplish for resolving the eigenvalue problem of completely continuous operators. Again let H be a completely continuous and self-adjoint operator. It follows from the theorem presented in (10.61) that the identity operator I is decomposed by[22]

$$I = \sum_{n=1}^{\infty} P_n, \tag{10.64}$$

---

[21] Particularly when 0 is also an eigenvalue of H, $\dim E(\lambda_\infty) = \infty$ and thus the representation of (10.62) is replaced by $\left\{ e_1^{(\infty)}, e_2^{(\infty)}, \ldots, e_k^{(\infty)} \right\}$. But for simplicity, we will consider only the case where 0 is excluded from the set of eigenvalues of H.

[22] See (I.16) for the finite-dimensional counterpart of the relation (10.64).

where P_n is the **projection operator** with respect to the eigenspace $E(\lambda_n)$. In addition, the theorem tells us that

$$H = \sum_{n=1}^{\infty} \lambda_n P_n, \tag{10.65}$$

which implies

$$Hx = \sum_{n=1}^{\infty} \lambda_n P_n x, \tag{10.66}$$

and thus

$$\|Hx\|^2 = \sum_{n=1}^{\infty} |\lambda_n|^2 \cdot \|P_n x\|^2. \tag{10.67}$$

Furthermore, (10.64) means that

$$y = \sum_{n=1}^{\infty} P_n y. \tag{10.68}$$

Combining the results, we obtain[23]

$$(Hx, y) = \sum_{n=1}^{\infty} \sum_{m=1}^{\infty} \lambda_m (P_m x, P_n y) = \sum_{n=1}^{\infty} \lambda_m (P_n x, P_n y). \tag{10.69}$$

Meanwhile, the orthonormal basis of the eigenspace $E(\lambda_n)$ is denoted by $\left\{e_1^{(n)}, e_2^{(n)}, \dots, e_{k_n}^{(n)}\right\}$. Since

$$P_n x = \sum_{i=1}^{k_n} (x, e_i^{(n)}) e_i^{(n)}, \tag{10.70}$$

(10.66) is rewritten by

$$Hx = \sum_{n=1}^{\infty} \sum_{i=1}^{k_n} \lambda_n (x, e_i^{(n)}) e_i^{(n)} \tag{10.71}$$

The expansion given by (10.71) is our main conclusion as to the eigenvalue problem for completely continuous operators H. Given an element x in a Hilbert space Ψ, its image Hx through a completely continuous (and self-adjoint) operator is represented by the double sum with respect to the labels n and i. Here, the label n indicates the n-th eigenspace $E(\lambda_n)$ which is a subspace of Ψ, and i indicates the ith eigenvector $e_i^{(n)}$ that lies within the n-th eigenspace.

Similar discussions on (10.67) and (10.69) yields

$$\|Hx\|^2 = \sum_{n=1}^{\infty} \sum_{i=1}^{k_n} \lambda_n^2 \left| \left(x, e_i^{(n)}\right) \right|^2, \tag{10.72}$$

and

$$(Hx, y) = \sum_{n=1}^{\infty} \sum_{i=1}^{k_n} \lambda_n \left(x, e_i^{(n)}\right) \overline{\left(y, e_i^{(n)}\right)}. \tag{10.73}$$

[23]In the derivation from (10.66) to (10.69), we can disregard the possibility that zero is one of the eigenvalue of H. Because the rightmost term in (10.69) is invariant even when $\lambda_\infty = 0$.

10.5 Spectrum of "continuous" operator

10.5.1 Consequence of loss of complete continuity

The discussion so far has shown the decomposition of a given element in a Hilbert space into the set of eigenvectors of its eigenspace. It is a natural extension of the finite-dimensional case, in which a vector is decomposed into a set of finite number of eigenvectors in the space considered. However, the success of the procedure totally relies on that the relevant operator is completely continuous and self-adjoint. It cannot be overemphasized that, if either of the two properties is absent from, the decomposition procedure fails in infinite dimension. So what occurs in the case of operators that is *not* completely continuous? Let us have a closer look on this issue in the argument below.

To address the issue, we consider the Hilbert space $L^2(\Omega)$ defined on the unit-length closed interval $\Omega = [0,1]$. Elements in $L^2(\Omega)$ are assumed to be complex-valued functions. Under these assumptions, any $f(t) \in L^2(\Omega)$ satisfies[24] $tf(t) \in L^2(\Omega)$ with $t \in \Omega$. Accordingly, the operator H defined by

$$H : f(t) \to tf(t), \tag{10.74}$$

is regarded as an operator that acts on $L^2(\Omega)$. The operator H we have introduced is known to be linear and bounded.[25] Furthermore, it is self-adjoint as proved by

$$(Hf,g) = \int_0^1 tf(t)\cdot\overline{g(t)}dt = \int_0^1 f(t)\cdot\overline{tg(t)}dt = (f,Hg). \tag{10.75}$$

Nevertheless, our operator H turns out to be never completely continuous. Therefore, this H prevents us from the decomposition procedure demonstrated in Section 10.4.3.

The absence of complete continuity in H is accounted for through the fact that the identity operator I is not completely continuous. Remind that H is an operator of I multiplied by a scalar t; namely, it is written by tI with the domain of $\Omega = [0,1]$. Since I is not completely continuous, therefore, H is not, too.[26]

A remarkable consequence from the loss of complete continuity in the operator H is that no real number λ can be an eigenvalue of H. This is clarified by proof of contradiction. Suppose tentatively that a real number λ satisfies

$$H\varphi = \lambda\varphi, \tag{10.76}$$

[24] This is proved by $\int_0^1 |tf(t)|^2 dt \le \int_0^1 |f(t)|^2 dt = \|f\|^2$, in which the inequality $|tf(t)| \le |f(t)|$ was used.

[25] The linearity of H is obvious. The boundedness is proved from $\|Hf\|^2 = \int_0^1 |tf(t)|^2 dt \le \|f\|^2$, which implies $\|Hf\| \le \|f\|$.

[26] The loss of complete continuity in H becomes clearer if we shift the domain from $[0,1]$ to $[1,2]$. Set an orthonormal basis of $L^2[1,2]$ by $\{e_n(t)\}$ $(n = 1,2,\ldots)$. Through the operator $H : f(t) \to tf(t)$, the basis vectors are mapped onto $te_n(t)$. Since $1 \le t \le 2$, we have

$$\|te_m(t) - te_n(t)\|^2 \ge \|e_m(t) - e_n(t)\|^2 = 2,$$

for any pairs of m and $n \ne m$. It is thus impossible to extract a convergent subsequence from $\{te_n(t)\}$.

for a certain $\varphi \in L^2[0,1]$. Then, we have

$$t\varphi(t) = \lambda\varphi(t) \tag{10.77}$$

almost everywhere in $0 \leq t \leq 1$. This implies that $\varphi(t) = 0$ almost everywhere, and thus φ should be the zero element in $L^2[0,1]$. Accordingly, λ is disable to be an eigenvalue of H.

10.5.2 Approximate eigenvector of continuous operator

We have observed that the continuous operator (not completely continuous) H defined by (10.74) cannot possess eigenvectors.

Interestingly, however, those λ lying at $0 \leq \lambda \leq 1$ behave in a similar manner to the true eigenvalues of H. To see it, we assume $\lambda > 0$ for convenience. Although there is no $\varphi(\neq 0)$ that exactly satisfies $H\varphi = \lambda\varphi$, we can find $\varphi(\neq 0)$ that satisfies

$$\|H\varphi - \lambda\varphi\| \leq \varepsilon\|\varphi\|, \tag{10.78}$$

no matter how small ε is. In other words, it is possible to make $H\varphi$ be nearly (not exactly) identical to $\lambda\varphi$, within the margin of error of $\varepsilon\|\varphi\|$. One example is a function $\varphi \in L^2[0,1]$ defined by

$$\varphi(t) \begin{cases} \neq 0, & \lambda - \varepsilon \leq t \leq \lambda \\ = 0. & \text{otherwise} \end{cases} \tag{10.79}$$

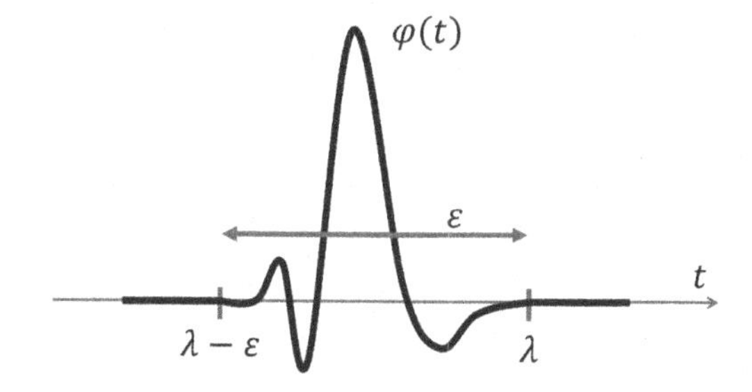

Figure 10.11: Profile of an example of a function $\varphi(t)$ that deviates from zero only within a interval $[\lambda - \varepsilon, \lambda]$.

Since $t\varphi(t) = 0$ at $t \notin [\lambda - \varepsilon, \lambda]$, we have

$$\begin{aligned} \|H\varphi - \lambda\varphi\|^2 &= \int_{\lambda-\varepsilon}^{\lambda} (t-\lambda)^2 |\varphi(t)|^2 dt \\ &< \varepsilon^2 \int_{\lambda-\varepsilon}^{\lambda} |\varphi(t)|^2 dt \\ &= \varepsilon^2 \|\varphi\|^2. \end{aligned} \tag{10.80}$$

Namely, the spiky function $\varphi(t)$ that deviates from zero only at the vicinity of $t = \lambda$ behaves effectively as the eigenfunction of the continuous operator $H : f(t) \in L^2[0,1] \to tf(t) \in L^2[0,1]$.

10.5.3 Set of continually changing projection operators

Our attempt to introduce the concept of "approximate eigenvector", which we denote by φ, is justified by the following argument based on the notion of continually changing eigenspaces and continually changing projection operators. To begin with, we set

$$E(\lambda) = \left\{ f \in L^2[0,1], t > \lambda \Rightarrow f(t) = 0 \text{ almost everywhere.} \right\} \tag{10.81}$$

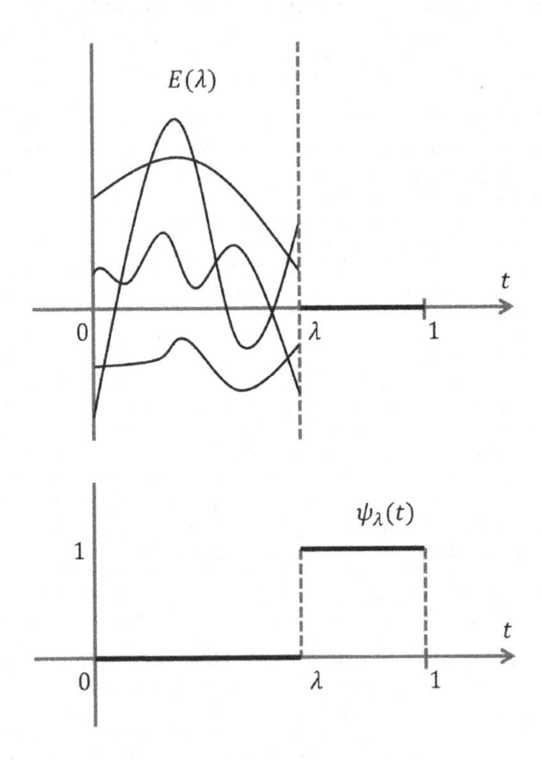

Figure 10.12: Top: A set of functions that are contained in the eigenspace $E(\lambda)$. Bottom: Stepwise function $\psi_\lambda(t)$ with a discontinuous jump at λ.

Obviously $E(\lambda)$ is a subspace of $L^2[0,1]$. In addition, $E(\lambda)$ is a closed space.[27] Suppose that λ increases from 0 to 1 in a continuous manner. The increase in λ causes a gradual change in the eigenspace $E(\lambda)$. All the eigenspaces produced by the continual increase in λ results in a set of closed subspaces, which denote by $\{E(\lambda)\}_{\lambda\in[0,1]}$. The elements of the set $\{E(\lambda)\}_{\lambda\in[0,1]}$ yields

$$\{0\} = E(0) \subset \cdots \subset E(\mu) \subset \cdots \subset E(\lambda) \subset \cdots \subset E(1) = L^2[0,1], \qquad (10.82)$$

where $\mu < \lambda$ is assumed. Figure 10.13 gives a schematic illustration of this inclusion relationship. Observe that in the figure, f belongs to $E(\lambda)$, but it does not belong to $E(\mu)$.

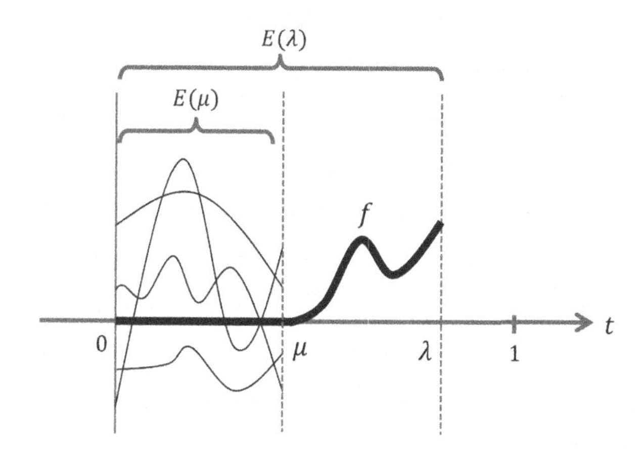

Figure 10.13: Diagram of the inclusion relation of the set of eigenspaces $\{E(\lambda)\}_{\lambda\in[0,1]}$. The inequality $\mu < \lambda$ implies $E(\mu) \subset E(\lambda)$ and thus there is a function f that belongs to $E(\lambda)$ but is excluded from $E(\mu)$.

The inclusion relation of $\{E(\lambda)\}_{\lambda\in[0,1]}$, given by (10.82), implies that $E(\lambda)$ is decomposed into the two subspaces: the one is the space $E^{\perp}(\lambda)$ that is orthogonal to $E(\lambda)$, and the other is $E(\mu)$. This decomposition is represented by

$$E(\lambda) = E(\mu) \oplus E^{\perp}(\lambda). \qquad (10.83)$$

---

[27]The closedness of $E(\lambda)$ is proved by setting the stepwise function:

$$\psi_\lambda(t) = \begin{cases} 0, & 0 \le t \le \lambda \\ 1. & t > \lambda \end{cases}$$

Suppose that $f_n \in E(\lambda)$ converges to f at $n \to \infty$. Then we have

$$\|\psi_\lambda f_n - \psi_\lambda f\| \le \|f_n - f\| \to 0. \ (n \to \infty)$$

In addition, $\psi_\lambda f_n = 0$ almost everywhere, as readily seen from Fig. 10.12. This means that $\psi_\lambda f = 0$ almost everywhere. Hence, we obtain the conclusion of $f \in E(\lambda)$. Since $f_n \in E(\lambda)$ implies $f \in E(\lambda)$, $E(\lambda)$ should be closed.

Meanwhile, the projection operator onto $E^{\perp}(\lambda)$ is formally written by

$$P(\lambda) - P(\mu). \tag{10.84}$$

Then, the projection operators of each sides of (10.83) can be equated as

$$P(\lambda) = P(\mu) + [P(\lambda) - P(\mu)]. \tag{10.85}$$

Acting the both sides of (10.85) onto $f \in L^2[0,1]$, we obtain

$$P(\lambda)f = P(\mu)f + [P(\lambda) - P(\mu)]f. \tag{10.86}$$

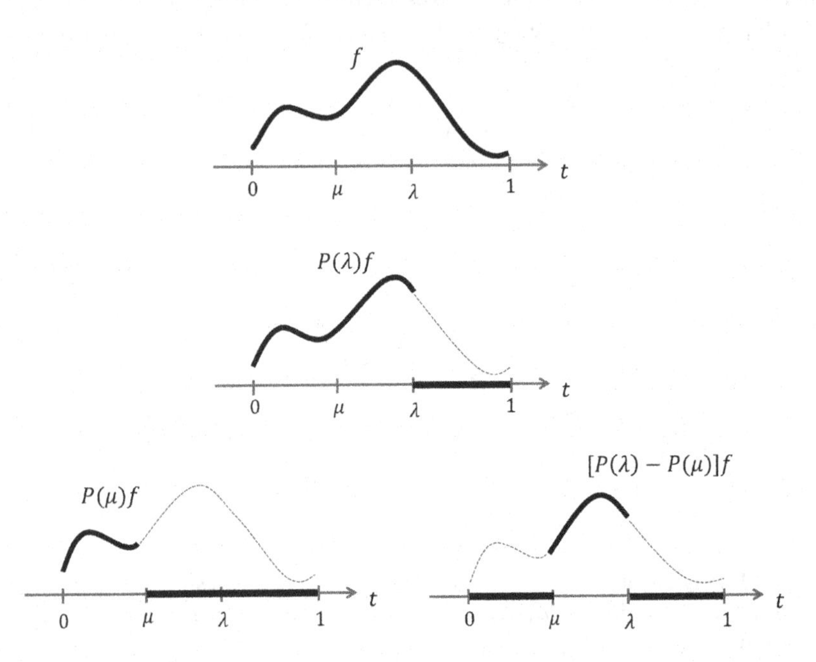

Figure 10.14: Schematic illustration of the decomposition of $P(\lambda)f$ into the sum of $P(\mu)f$ and $[P(\lambda) - P(\mu)]f$ represented by (10.86).

Geometric implication of (10.86) is plotted in Fig.10.14. If we set

$$\varphi = [P(\lambda) - P(\mu)]f, \tag{10.87}$$

a parallel discussion that led us to (10.80) will yield

$$\|H\varphi - \lambda\varphi\| \le (\lambda - \mu)\|\varphi\|. \tag{10.88}$$

Namely, the "spiky" function φ, which is produced by operating $P(\lambda) - P(\mu)$ on f, behaves as an approximate eigenfunction for the effective eigenvalue λ. To improve the accuracy of approximation, we have only to diminish the value of $\lambda - \mu$ in (10.88), or equivalently, to set μ closer to λ. Since the scalar μ can be tuned in a continuous manner, it is expected that an ultimate accuracy is attained by taking the limit of $\mu \to \lambda$. Is this conjecture true? The positive answer will be given in the next subsection.

10.5.4 Spectral decomposition of operator

Look at Fig. 10.15. We introduce a point sequence $\{\lambda_k\}$ on the closed interval of [0,1] such that

$$0 = \lambda_0 < \lambda_1 < \cdots < \lambda_{k-1} < \lambda_k < \cdots < \lambda_n = 1. \tag{10.89}$$

Using the sequence of projection operators $\{P(\lambda_k)\}$, a given function f is decomposed to a set of spiky functions. Each of the spiky functions is written by

$$[P(\lambda_k) - P(\lambda_{k-1})] f = \begin{cases} f, & \text{within } [\lambda_{k-1}, \lambda_k] \\ 0. & \text{otherwise} \end{cases} \tag{10.90}$$

Summing up all the functions of $[P(\lambda_k) - P(\lambda_{k-1})] f$ from $k = 0$ to $k = n$, the original function f is reproduced as

$$f \simeq \sum_{k=1}^{n} [P(\lambda_k) - P(\lambda_{k-1})] f. \tag{10.91}$$

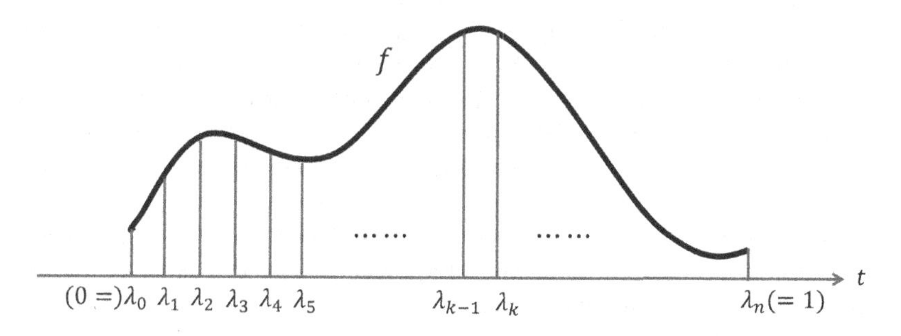

Figure 10.15: Decomposition of f into a set of functions, $[P(\lambda_k) - P(\lambda_{k-1})] f$, from $k = 0$ to $k = n$.

Note that (10.91) implies

$$Hf \simeq \sum_{k=1}^{n} H [P(\lambda_k) - P(\lambda_{k-1})] f. \tag{10.92}$$

If we take the value $\lambda_{k-1} - \lambda_k$ be sufficiently small, the operator $H [P(\lambda_k) - P(\lambda_{k-1})]$ that appears in the right side of (10.92) becomes close enough to $\lambda_k [P(\lambda_k) - P(\lambda_{k-1})]$. This fact motivates us to write tentatively by

$$H \simeq \sum_{k=1}^{n} \lambda_k [P(\lambda_k) - P(\lambda_{k-1})] \;\to_?\; \int \lambda dP(\lambda). \tag{10.93}$$

The last expression in (10.93) owes to the analogy with elementary integration. It is based on the conjecture that the finite sum in the middle of (10.93) is reduced to the integral under a certain limiting procedure. Namely, we would like to segmentalize the interval $[\lambda_{k-1}, \lambda_k]$ very finely and take the limit of $\lambda_{k-1} - \lambda_k \to 0$. It is known that the formal integral representation presented in (10.93) is meaningful for any continuous and self-adjoint operator acting on Hilbert spaces. The results are summarized by the following theorem.

Theorem (Spectral decomposition):[28]
Let H be a continuous and self-adjoint operator. The H is described using an appropriate set of the projection operators $\{P(\lambda)\}_{\lambda \in \mathbb{R}}$ by

$$H = \int \lambda dP(\lambda). \tag{10.94}$$

Here, the set $\{P(\lambda)\}_{\lambda \in \mathbb{R}}$ satisfies the condition that

$$\lambda \leq a \Rightarrow P(\lambda) = 0 \text{ and } \lambda \geq b \Rightarrow P(\lambda) = I, \tag{10.95}$$

for a pair of $a, b \in \mathbb{R}$. ■

The real number λ, which appears in (10.94), is called a **continuous spectrum** of the operator H. Importantly, λ can move continuously within $0 \leq \lambda \leq 1$. The continuity of λ is in contrast with the discreteness of eigenvalues of finite-dimensional matrices that we encounter in elementary linear algebra. As a result, we have successfully established the new mathematical entities: the approximated eigenvalues λ_k and the approximated eigenspaces $E(\lambda_k)$, both of which are peculiar to continuous (not completely continuous) and self-adjoint operators.

Compared with the case of completely continuous operators, there are many conceptual discrepancies from the elementary linear algebra that is based on the manipulation of finite-dimensional matrices.

10.6 Practical importance of non-continuous operators

10.6.1 "Non-commuting" means "non-continuous"

We conclude this chapter by surveying the relevance of mathematical operators to the theoretical framework of quantum physics. Quantum mechanics theory is based on the fundamental relation between two specific operators, Q and P, which can be written as

$$PQ - QP = \frac{h}{2\pi i} I. \tag{10.96}$$

Here, Q is the **position operator** that corresponds to an observable position of a quantum particle, P is the **momentum operator**, and h is a physical constant called **Plank's constant**. Using the terminology of quantum mechanics, the relation in

---

[28]The proof of the theorem is demonstrated, for instance, in Book "*Functional analysis*" written by Frigyes Riesz and Béla Sz.-Nagy (Dover Publications, 1990).

(10.96) is called the **uncertainty principle**. It clearly implies that these two operators are non-commutative, which can be expressed using

$$PQ \neq QP. \tag{10.97}$$

In the following discussion, we address the mathematical consequences of operator non-commutativity. Surprisingly, we see that these operators, which are mutually non-commuting, should not be continuous.[29]

The first consequence of non-commutativity is that the fundamental relation in (10.96) fails given a pair of finite-dimensional matrices. To show it, we tentatively assume a pair of n-dimensional matrices, $\boldsymbol{A}$ and $\boldsymbol{B}$, satisfies the following relation.

$$\boldsymbol{AB} - \boldsymbol{BA} = \boldsymbol{I}. \tag{10.98}$$

If we take the **trace**[30] of both sides, we will obtain the following result.

$$\mathrm{Tr}(\boldsymbol{AB}) - \mathrm{Tr}(\boldsymbol{BA}) \stackrel{??}{=} \mathrm{Tr}(\boldsymbol{I}). \tag{10.99}$$

The left side of (10.99) equals zero because the matrices in the trace of a product can be switched: $\mathrm{Tr}(\boldsymbol{AB}) = \mathrm{Tr}(\boldsymbol{BA})$. But the right side of (10.99) is clearly n. Therefore, (10.99) fails in the case of finite-dimensional matrices.

This conclusion implies that, when establishing the theory of quantum mechanics, operators that act on infinite-dimensional spaces must be used. Even in this case, not all operators are effective. In fact, only the class of *non*-continuous operators should be considered in order to ensure that the fundamental relation in (10.96) is valid. This argument is stated in the following theorem.

Theorem:
Let A and B be continuous (and thus, bounded) operators that act on infinite-dimensional Hilbert spaces. Any two given operators, A and B, will not satisfy the following relation.

$$AB - BA = I. \quad \blacksquare \tag{10.100}$$

Proof:
This is a proof by contradiction. Assume that (10.100) holds for certain bounded operators A and B. Clearly, we have $A \neq 0$ and $B \neq 0$. If B acts from the right on both sides of (10.100), we obtain

$$AB^2 - BAB = B. \tag{10.101}$$

---

[29]In this context, the theory of non-commutative operators in Hilbert spaces describes the nature of quantum mechanical particles.

[30]In linear algebra, the **trace** of an n-by-n square matrix, $\boldsymbol{A}$, is defined to be the sum of the elements on the diagonal from the upper left to the lower right of $\boldsymbol{A}$, i.e., $\mathrm{Tr}(\boldsymbol{A}) = a_{11} + a_{22} + \cdots + a_{nn} = \sum_{i=1}^{n} a_{ii}$. Here, a_{nn} denotes the entry in the n-th row and the n-th column of $\boldsymbol{A}$. The trace of a matrix is the sum of the (complex) eigenvalues, and it is invariant with respect to a change of basis.

Because $AB = BA + I$, we know

$$AB^2 - B^2A = 2B. \tag{10.102}$$

Similarly, if B acts from the right on both sides, and we use relation $AB = BA + I$, we can conclude that

$$AB^3 - B^3A = 3B^2. \tag{10.103}$$

Repeating this procedure produces the following two relations:

$$AB^{n-1} - B^{n-1}A = nB^{n-2}, \tag{10.104}$$

and

$$AB^n - B^nA = nB^{n-1}. \tag{10.105}$$

It follows from (10.104) that if $B^{n-1} = 0$, then $B^{n-2} = 0$. Similarly, if $B^{n-2} = 0$, then $B^{n-3} = 0$. Namely, the condition of $B^{n-1} = 0$ implies $B^{n-2} = \cdots = B^2 = B = 0$, which contradicts our original assumption that $B \neq 0$. Hence, B^{n-1} must not be a zero operator.

Now if we consider (10.105), it follows that

$$\begin{aligned} n\|B^{n-1}\| &\leq \|AB^n\| + \|B^nA\| \\ &\leq \|AB\| \cdot \|B^{n-1}\| + \|BA\| \cdot \|B^{n-1}\| \\ &= (\|AB\| + \|BA\|) \cdot \|B^{n-1}\|. \end{aligned} \tag{10.106}$$

Because $B^{n-1} \neq 0$, we can divide the leftmost and rightmost sides of (10.106) by $\|B^{n-1}\|$ and obtain

$$n \leq \|AB\| + \|BA\|. \tag{10.107}$$

If we recall that n is any arbitrary natural number, the value of n can be as large as we like. However, this contradicts the boundedness of AB and BA. As a consequence, we can say that no pair of bounded operators, A and B, satisfies (10.100). ■

Keypoint: There is no continuous operator that satisfies the uncertainty principle in quantum physics.

10.6.2 *Ubiquity of non-commuting operators*

As we just learned, the uncertainty principle, which plays a central role in the theory of quantum mechanics, is a relation between non-continuous operators. Because of the non-continuity of the operators, we can no longer rely on a conceptual analogy to elementary linear algebra in order to capture the mathematical properties of physical observables (i.e., non-continuous operators) within the field of quantum mechanics. Even the most essential principle in physics requires a rather complicated, advanced class of operators beyond the simple, well-behaved operators (continuous or completely continuous operators). Therefore, while exploring concepts in quantum mechanical theory, we should abandon intuitive understanding and construct a purely logical thought process in order to describe the mathematical properties of non-continuous operators. This is a primary reason the study of functional analysis benefits those involved in physics and related engineering fields.

It is also interesting to note that non-continuous operators frequently appear in the mathematical arguments of the function spaces of $L^2(\mathbb{R})$. For example, if we suppose operator Q acts on $f \in L^2(\mathbb{R})$, we can define it using

$$Qf(t) = tf(t). \tag{10.108}$$

This operator, Q, is not continuous. We can prove this non-continuity by considering a sequence of functions, g_n, defined by

$$g_n(t) = \begin{cases} \frac{1}{t}, & 1 \le t \le n+1 \\ 0, & t < 1,\ t > n+1 \end{cases}, \tag{10.109}$$

for $n = 1, 2, \cdots$. From this, it follows that

$$\|g_n\| \le 1 \ (n = 1, 2, \cdots), \tag{10.110}$$

but that

$$\|Qg_n\| \to \infty \ (n \to \infty). \tag{10.111}$$

In other words, a simple scalar-multiplication operator, Q, that acts on $f \in L^2(\mathbb{R})$ will generally not be continuous[31] despite the ubiquity and simplicity of its representation.

Another important non-continuous operator is the differential operator P that acts on $f \in C_0(\mathbb{R})$:

$$Pf(t) = \frac{d}{dt} f(t). \tag{10.112}$$

Given the relation in (10.112), operator P maps all elements in the function space of $C_0(\mathbb{R})$ onto images contained in the space of $L^2(\mathbb{R})$.

The non-continuity of P is proved using a function sequence, $\{f_n\}$, in $C_0(\mathbb{R})$ that yields $\|f_n\| \to 0$ when $n \to \infty$, but $\|Pf_n\| \to \infty$. Figure 10.16 illustrates an example of this sequence, $\{f_n\}$. The f_n depicted in this figure are elements of $C_0(\mathbb{R})$. A straightforward calculation yields

$$\|f_n\|^2 = 2 \int_0^{1/n^3} (-n^4 t + n)^2 dt = \frac{2}{3} \cdot \frac{1}{n}, \tag{10.113}$$

which implies that

$$\|f_n\| \to 0 \ (n \to \infty). \tag{10.114}$$

[31] The situation drastically changes if Q acts on $f \in L^2[0,1]$ instead of on $f \in L^2(\mathbb{R})$. In the former case, Q may be continuous, as is demonstrated in section 10.5.1.

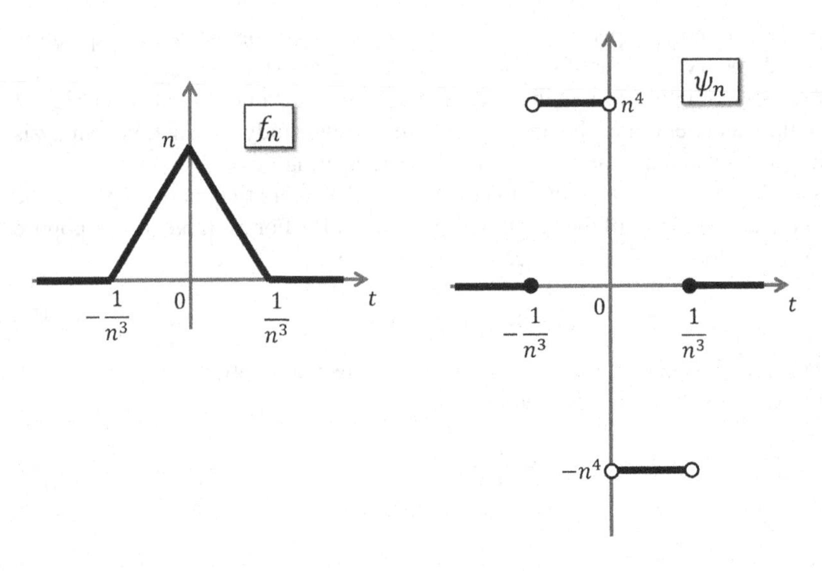

Figure 10.16: Profiles of $f_n \in C_0(\mathbb{R})$ and $\psi_n = Pf_n \in L^2(\mathbb{R})$.

In addition, we can introduce $\psi_n \equiv Pf_n$, which describes the slope of curve $f_n(t)$, except when $t = \pm 1/n^3$ and $t = 0$. Functions $\psi_n(t)$ are defined almost everywhere. Thus, $\psi_n \in L^2(\mathbb{R})$, and it readily follows that

$$\|\psi_n\|^2 = 2n. \tag{10.115}$$

Therefore, $\|\psi_n\| \to \infty$ at the limit of $n \to \infty$. The disagreement between $\lim_{n\to\infty} \|f_n\|$ and $\lim_{n\to\infty} \|\psi_n\|$ indicates that differential operator P is not continuous on its domain, $C_0(\mathbb{R})$, which is a subspace of $L^2(\mathbb{R})$.

Recall that operators P and Q are both non-continuous. These two operators satisfy the non-commuting relation,

$$PQ - QP = I, \tag{10.116}$$

which extends beyond finite-dimensional matrices. In fact, for $f \in C_0(\mathbb{R})$, we have

$$\begin{aligned} (PQ - QP) f(t) &= \frac{d}{dt}\Big(tf(t)\Big) - t\frac{df(t)}{dt} \\ &= f(t) + t\frac{df(t)}{dt} - t\frac{df(t)}{dt} = f(t). \end{aligned} \tag{10.117}$$

In addition to the position and momentum operators, many non-continuous operators satisfy the uncertainty principle. Hence, we need an in-depth understanding of the properties of non-continuous operators when exploring the mathematical details of quantum physics. This is a primary reason non-mathematics readers benefit from the study of functional analysis.

Appendix A

Real Number Sequence

A.1 Convergence of real sequence

The Appendix A describes fundamental definitions and notions associated with **sequences of real numbers** (called a **real sequence**).[1]

We start with a precise definition of convergence of a real sequence. In fact, it serves as an initial and crucial step into various branches in mathematics.

Definition (Convergence of a real sequence):
A real sequence $\{x_n\}$ is said to be **convergent** if there exists a real number x with the following property: For every $\varepsilon > 0$, there is an integer N such that

$$n \geq N \Rightarrow |x_n - x| < \varepsilon. \quad \blacksquare \tag{A.1}$$

Attention should be paid to the arbitrary property in the magnitude of ε. No matter how small ε we choose, it must always be possible to find a number N that satisfies the condition of (A.1). Consequently, the value of N that we need will increase as ε decreases.

When $\{x_n\}$ is convergent, the number x specified in this definition is called a **limit** of the sequence $\{x_n\}$, and we say that x_n converges to x. This is expressed symbolically by writing

$$\lim_{n \to \infty} x_n = x, \tag{A.2}$$

[1]Keep in mind the difference between a sequence and a set. A sequence $\{x_n\}$ is an *ordered* list of numbers, in which same numbers may be repeated. A set $\{x_n\}$ is merely the defining *range* of the variables x_n. For instance, the sequence $\{1, 1, 1, \cdots\}$ is a constant sequence in which every term is equal to 1, while the set $\{1\}$ contains only one element.

or simply by

$$x_n \to x \text{ at } n \to \infty. \tag{A.3}$$

If $\{x_n\}$ is not convergent, it is called **divergent**.

It should be emphasized that only a restricted class of convergent sequences includes the limit in the same sequence. Given below is an example in which $x = \lim_{n\to\infty} x_n$ but $x \neq x_n$ for any n.

Example:
Consider an infinite sequence $\{x_n\}$ defined by

$$\{x_n\} = \{3.1,\ 3.14,\ 3.142,\ \cdots,\ x_n,\ \cdots\}, \tag{A.4}$$

in which $x_n \in \mathbb{Q}$ is a rational number to n places of decimals close to π. Since the difference $|x_n - \pi|$ is less than 10^{-n}, it is possible for any $\varepsilon > 0$ to find an N such that

$$n \geq N \ \Rightarrow \ |x_n - \pi| < \varepsilon. \tag{A.5}$$

This means that

$$\lim_{n\to\infty} x_n = \pi. \tag{A.6}$$

However, the limit (i.e., π) is an irrational number, thus not being contained in the sequence $\{x_n\}$. ■

Keypoint: The limit x of a sequence $\{x_n\}$ may and may not belong to $\{x_n\}$.

A.2 Bounded sequence

Together with the convergence property, there is another fundamental property of real sequences, called the boundedness.

Definition (Bounded sequence):
A real sequence $\{x_n\}$ is said to be **bounded** if there is a positive number M such that

$$|x_n| \leq M \ \text{ for all } \ n \in \mathbb{N}. \ \blacksquare \tag{A.7}$$

The following is an important relation between convergence and boundedness of a real sequence.

Theorem:
If a sequence is convergent, then it is bounded. ■

Proof:
Suppose $x_n \to x$. If we choose $\varepsilon = 1$ in (A.1), there exists an integer N such that

$$|x_n - x| < 1 \text{ for all } n \geq N. \tag{A.8}$$

Since $|x_n| - |x| \leq |x_n - x|$, it follows that

$$|x_n| < 1 + |x| \text{ for all } n \geq N. \tag{A.9}$$

Setting $M = \max\{|x_1|, |x_2|, \cdots, |x_{N-1}|, 1 + |x|\}$ yields

$$|x_n| < M \text{ for all } n \in \mathbb{N}, \tag{A.10}$$

which means that $\{x_n\}$ is bounded. ■

Observe that the converse of the theorem is false. Namely, even if a sequence is bounded, it may or may not be convergent. In fact, the sequence

$$\{1, -1, 1, -1, \cdots, (-1)^n, \cdots\} \tag{A.11}$$

is divergent, though it is bounded.

Keypoint:
If a sequence is convergent, then it *must* be bounded.
But even if a sequence is bounded, it *may or may not* be convergent.

A.3 Uniqueness of the limit of real sequence

The limit of a convergent sequence is always determined uniquely.

This fact is confirmed by proof by contradiction as follows. Assume that a sequence $\{x_n\}$ has two different limits written by

$$x = \lim_{n\to\infty} x_n \text{ and } y = \lim_{n\to\infty} x_n, \tag{A.12}$$

with the assumption that $x \neq y$. Then we can find a neighborhood V_1 of x and a neighborhood V_2 of y such that

$$V_1 \cap V_2 = \emptyset. \tag{A.13}$$

For example, take $V_1 = (x - \varepsilon, x + \varepsilon)$ and $V_2 = (y - \varepsilon, y + \varepsilon)$, where

$$\varepsilon = \frac{|x - y|}{2}. \tag{A.14}$$

Since $x_n \to x$, all but a finite number of terms of the sequence lie in V_1. Similarly, since $y_n \to y$, all but a finite number of its terms also lie in V_2. However, these results contradict to (A.13). As a consequence, we conclude that the limit of a sequence should be unique.

Keypoint: A convergent sequence has no more than one limit.

Appendix B

Cauchy Sequence

B.1 What is Cauchy sequence?

In the main text of this book, we have repeatedly observed that the concept of Cauchy sequence plays a pivotal role in formulating many theorems of functional analysis. We revisit again the precise definition.

Definition (Cauchy sequence):
The sequence $\{x_n\}$ is called a **Cauchy sequence**[1] if, for every positive number ε, there is a positive integer N such that

$$m, n > N \Rightarrow |x_n - x_m| < \varepsilon. \quad \blacksquare \tag{B.1}$$

It is stated above that in every Cauchy sequence, the terms become as close to each other as we choose. This feature of Cauchy sequences is expected to hold for any convergent sequence, since the terms of a convergent sequence have to approach each other as they approach a common limit. This conjecture is assured in part by the following theorem.

Theorem:
If a sequence $\{x_n\}$ is convergent, then it is a Cauchy sequence. ■

[1]Cauchy sequence is also known as **fundamental sequence**.

Proof:
Suppose $\lim_{n\to\infty} x_n = x$ and ε is any positive number. From the hypothesis, there exists a positive integer N such that

$$n > N \Rightarrow |x_n - x| < \frac{\varepsilon}{2}. \tag{B.2}$$

Now if we take $m, n \geq N$, then

$$|x_n - x| < \frac{\varepsilon}{2}, \text{ and } |x_m - x| < \frac{\varepsilon}{2}. \tag{B.3}$$

It thus follows that

$$|x_n - x_m| \leq |x_m - x| + |x_n - x| < \varepsilon, \tag{B.4}$$

which means that $\{x_n\}$ is a Cauchy sequence. ■

This theorem naturally arise the following question: Is the converse true? In other words, we would like to know whether all Cauchy sequences are convergent or not. The answer to this question is the very what the Cauchy criterion states, as proven in the next argument.

B.2 Cauchy criterion for real number sequence

The following is one of the fundamental theorem for the study of functional analysis.

> **Theorem (Cauchy criterion):**
> A sequence of real numbers is convergent if and only if it is a Cauchy sequence. ■

Therefore, convergence of a given sequence can be examined by seeing whether it is a Cauchy sequence or not. Realize that the validity of this criterion was partly proven by demonstrating the previous theorem (See §B.1). Hence, in order to complete the proof of the criterion, we need only prove that *every Cauchy sequence is convergent.*

For the aim, let $\{x_n\}$ be a Cauchy sequence, and let S be a set composed of the elements of the sequence, represented by $S = \{x_n : n \in \mathbb{N}\}$. We examine the following two cases by turns: i) the set S is finite, or ii) S is infinite.

i) We first consider the case in which the set S is finite. From the hypothesis, it follows that for a given $\varepsilon > 0$, there exists an integer N such that

$$m, n > N \Rightarrow |x_n - x_m| < \varepsilon. \tag{B.5}$$

Since S is finite, one of the terms in the sequence $\{x_n\}$, say x, should be repeated infinitely often in order to satisfy (B.5). This implies the existence of an $m (> N)$ such that $x_m = x$. Hence we have

$$n > N \Rightarrow |x_n - x| < \varepsilon, \tag{B.6}$$

which means that $x_n \to x$.

ii) Next we consider the case that S is infinite. It can be shown that every Cauchy sequence is bounded; See Appendix A.2. Hence, by virtue of **Bolzano-Weierstrass' theorem**,[2] the sequence $\{x_n\}$ necessarily has an accumulation point x. We shall prove that $\{x_n\}$ converges to the accumulation point. Given $\varepsilon > 0$, there is an integer N such that

$$m, n > N \Rightarrow |x_n - x_m| < \varepsilon. \tag{B.7}$$

From the definition of an accumulation point, the open interval $(x - \varepsilon, x + \varepsilon)$ must contain an infinite number of terms of the sequence $\{x_n\}$. Hence, there is an $m \geq N$ such that $x_m \in (x - \varepsilon, x + \varepsilon)$, i.e., such that $|x_n - x_m| < \varepsilon$. Now, if $n \geq N$, then

$$|x_n - x| \leq |x_n - x_m| + |x_m - x| < \varepsilon + \varepsilon = 2\varepsilon, \tag{B.8}$$

which proves $x_n \to x$.

The two results for the cases of i) and ii) indicate that *every Cauchy sequence (finite and infinite) is convergent.* Recall again that its converse, *every convergent sequence is a Cauchy sequence*, has been already proven in §B.1. Eventually, the proof of the Cauchy criterion has been completed.

Keypoint: A sequence is convergent $\Longleftrightarrow$ It is a Cauchy sequence.

[2]**Bolzano-Weierstrass' theorem** states that every infinite and bounded sequence of real numbers has at least one accumulation point in $\mathbb{R}$. The proof of the theorem is described, for instance, in Book: "*Higher Mathematics for Physics and Engineering*" by H. Shima and T. Nakayama (Pan Stanford Publishing, Singapore, 2010).

Appendix C

Real Number Series

C.1 Limit of real number series

The Appendix C focuses on convergence properties of **infinite series**. The importance of this issue should be obvious when we discuss the properties of Hilbert spaces and orthogonal polynomial expansions, in which infinite series of numbers or functions enter quite often.

To begin with, we briefly review the basic properties of infinite series of real numbers. Suppose an infinite sequence $\{a_1, a_2, \cdots, a_n, \cdots\}$ of real numbers. Then, we can form another infinite sequence $\{A_1, A_2, \cdots, A_n, \cdots\}$ with the definition:

$$A_n = \sum_{k=1}^{n} a_k. \tag{C.1}$$

Here, A_n is called the nth **partial sum** of the sequence $\{a_n\}$, and the corresponding infinite sequence $\{A_n\}$ is called the **sequence of partial sums** of $\{a_n\}$. The infinite sequence $\{A_n\}$ may and may not be convergent, which depends on the feature of $\{a_n\}$.

Let us introduce an **infinite series** defined by

$$\sum_{k=1}^{\infty} a_k = a_1 + a_2 + \cdots. \tag{C.2}$$

The infinite series (C.2) is said to converge if and only if the sequence $\{A_n\}$ converges to the limit denoted by A. In other words, the series (C.2) converges if and only if the **sequence of the remainder** $R_{n+1} = A - A_n$ converges to zero. When $\{A_n\}$ is convergent, its limit A is called the **sum of the infinite series** of (C.2), and it is possible to write

$$\sum_{k=1}^{\infty} a_k = \lim_{n\to\infty} \sum_{k=1}^{n} a_k = \lim_{n\to\infty} A_n = A. \tag{C.3}$$

Otherwise, the series (C.2) is said to diverge.

The limit of the sequence $\{A_n\}$ is formally defined in line with Cauchy's procedure as shown below.

Definition (Limit of a sequence of partial sums):
The sequence $\{A_n\}$ has a limit A if for any small $\varepsilon > 0$, there exists a number N such that

$$n > N \Rightarrow |A_n - A| < \varepsilon. \quad \blacksquare \tag{C.4}$$

Example 1:
The infinite series $\sum_{k=1}^{\infty}\left(\frac{1}{k}-\frac{1}{k+1}\right)$ converges to 1, because

$$A_n = \sum_{k=1}^{n}\left(\frac{1}{k}-\frac{1}{k+1}\right) = 1 - \frac{1}{n+1} \to 1. \quad (n\to\infty) \quad \blacksquare \tag{C.5}$$

Example 2:
The series $\sum_{k=1}^{\infty}(-1)^k$ diverges, because the sequence

$$A_n = \sum_{k=1}^{n}(-1)^k = \begin{cases} 0, & (n \text{ is even}) \\ -1, & (n \text{ is odd}) \end{cases} \tag{C.6}$$

approaches no limit. $\blacksquare$

Example 3:
The series $\sum_{k=1}^{\infty} 1 = 1+1+1+\cdots$ diverges, since the sequence $A_n = \sum_{k=1}^{n} 1 = n$ increases without limit as $n\to\infty$. $\blacksquare$

C.2 Cauchy criterion for real number series

Here is given a direct application of the Cauchy criterion to the sequence $\{A_n\}$ consisting of the partial sum $A_n = \sum_{k=1}^{n} a_k$.

Theorem (Cauchy criterion for infinite series):
The sequence of partial sums $\{A_n\}$ converges if and only if for any small $\varepsilon > 0$ there exists a number N such that

$$n, m > N \Rightarrow |A_n - A_m| < \varepsilon. \quad \blacksquare \tag{C.7}$$

The Cauchy criterion noted above provides a necessary and sufficient condition for convergence of the sequence $\{A_n\}$., similarly to the case of real sequences. Below is an important theorem associated with the convergent infinite series.

Theorem:
If an infinite series $\sum_{k=1}^{\infty} a_k$ is convergent, then

$$\lim_{k\to\infty} a_k = 0. \tag{C.8}$$

Proof:
From hypothesis, we have

$$\lim_{n\to\infty} \sum_{k=1}^{n} a_k = \lim_{n\to\infty} A_n = A, \quad \text{and} \quad a_n = A_n - A_{n-1}. \tag{C.9}$$

Hence,

$$\lim_{n\to\infty} a_n = \lim_{n\to\infty} (A_n - A_{n-1}) = A - A = 0. \quad \blacksquare \tag{C.10}$$

According to the theorem above, $\lim_{n\to\infty} a_n = 0$ is a necessary condition for the convergence of A_n. However, it is not a sufficient one, as shown in the following example.

Example:
Let $a_k = 1/\sqrt{k}$. Although $\lim_{k\to\infty} a_k = 0$, the corresponding infinite series $\sum a_k$ diverges as seen by

$$\begin{aligned} \sum_{k=1}^{n} a_k &= 1 + \frac{1}{\sqrt{2}} + \cdots + \frac{1}{\sqrt{n}} \\ &\geq \frac{1}{\sqrt{n}} + \frac{1}{\sqrt{n}} + \cdots + \frac{1}{\sqrt{n}} = \frac{n}{\sqrt{n}} = \sqrt{n} \to \infty. \quad \blacksquare \end{aligned}$$

Keypoint:

1. $\sum_{k=1}^{\infty} a_k$ is convergent $\Rightarrow \lim_{k\to\infty} a_k = 0$.
2. $\lim_{k\to\infty} a_k \neq 0 \Rightarrow \sum_{k=1}^{\infty} a_k$ is divergent.

Appendix D

Continuity and Smoothness of Function

D.1 Limit of function

Having learned the limit of sequences and series of real numbers, we turn our attention to the limit of functions. Let A be a real number and $f(x)$ a real-valued function of a real variable $x \in \mathbb{R}$. A formal notation of the above function is given by the mapping relation $f : \mathbb{R} \to \mathbb{R}$. The statement:

"*the limit of $f(x)$ at $x = a$ is A*"

means that the value of $f(x)$ can be as close to A as desired by setting x to be sufficiently close to a. This will be stated formally by the following definition.

Definition (Limit of a function):
A function $f(x)$ is said to have the **limit** A as $x \to a$ if and only if, for every $\varepsilon > 0$, there exists a number $\delta > 0$ such that

$$|x-a| < \delta \Rightarrow |f(x)-A| < \varepsilon. \quad \blacksquare \tag{D.1}$$

The limit of $f(x)$ is written symbolically by

$$\lim_{x\to a} f(x) = A,$$

or

$$f(x) \to A \ \text{ for } \ x \to a.$$

If the first inequalities in (D.1) are replaced by $0 < x - a < \delta$ (or $0 < a - x < \delta$), we say that $f(x)$ approaches A as $x \to a$ from above (or below) and write

$$\lim_{x \to a+} f(x) = A, \quad \left(\text{or} \lim_{x \to a-} f(x) = A \right).$$

This is called the **right-hand** (or **left-hand**) **limit** of $f(x)$. Or, these two are collectively called **one-sided limits**.

Below is given a necessary-sufficient condition for the existence of $\lim_{x \to a} f(x)$.

Theorem:
The limit of $f(x)$ at $x = a$ exists if and only if

$$\lim_{x \to a+} f(x) = \lim_{x \to a-} f(x). \quad \blacksquare \tag{D.2}$$

Proof:
If $\lim_{x \to a} f(x)$ exists and is equal to A, it readily follows that

$$\lim_{x \to a+} f(x) = \lim_{x \to a-} f(x) = A. \tag{D.3}$$

We now consider the converse. Assume that (D.2) holds. This obviously means that both the one-sided limit exist at $x = a$. Hence, given $\varepsilon > 0$, there are $\delta_1 > 0$ and $\delta_2 > 0$ such that

$$\begin{aligned} 0 < x - a < \delta_1 &\quad \Rightarrow \quad |f(x) - A| < \varepsilon, \\ 0 < a - x < \delta_2 &\quad \Rightarrow \quad |f(x) - A| < \varepsilon. \end{aligned}$$

Let $\delta = \min\{\delta_1, \delta_2\}$. If x satisfies $0 < |x - a| < \delta$, then either

$$0 < x - a < \delta \le \delta_1 \text{ or } 0 < a - x < \delta \le \delta_2.$$

In either case, we have $|f(x) - A| < \varepsilon$. That is, we have seen that for a given ε, there exists δ such that

$$0 < |x - a| < \delta \Rightarrow |f(x) - A| < \varepsilon.$$

Therefore we conclude that

$$\text{Equation (D.2) holds} \Rightarrow \lim_{x \to a} f(x) = A,$$

and the proof has been completed. $\blacksquare$

D.2 Continuity of function

In general, the value of $\lim_{x \to a} f(x)$ has nothing to do with the value of (and the existence of) $f(a)$. For instance, the function given by

$$f(x) = \begin{cases} 2 - x^2 & x \neq 1, \\ 2 & x = 1, \end{cases}$$

provides

$$\lim_{x \to 1} f(x) = 0 \text{ and } f(1) = 2,$$

which are difference quantitatively from each other. This mismatch occurring at $x = 1$ results in a lack of geographical continuity in the curve of the graph $y = f(x)$ as depicted in Fig. D.1. In mathematical language, continuity of the curve of $y = f(x)$ is accounted for by the following statement.

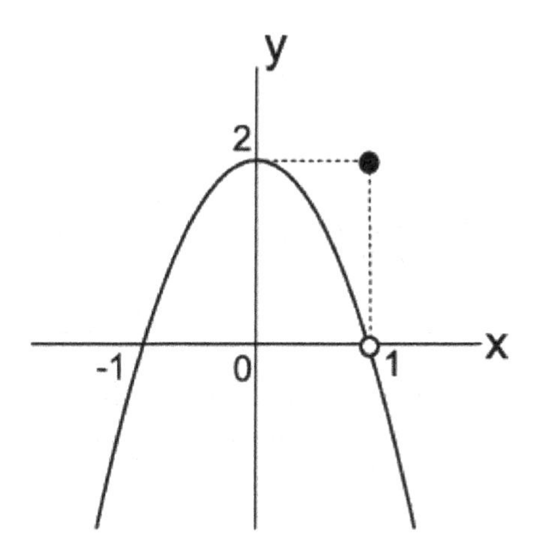

Figure D.1: A discontinuous function $y = f(x)$ at $x = 1$.

Definition (Continuous function):
The function $f(x)$ is said to be **continuous** at $x = a$ if and only if, for every $\varepsilon > 0$, there exists $\delta > 0$ such that

$$|x - a| < \delta \Rightarrow |f(x) - f(a)| < \varepsilon. \quad \blacksquare \tag{D.4}$$

The definition noted above is apparently similar to the definition of the limit of $f(x)$ at $x = a$ (See §D.1). However, there is a crucial difference between them. When considering the limit of $f(x)$ at $x = a$, we are only interested in the behavior of $f(x)$ *in the vicinity of* the point $x = a$, not *just at* $x = a$. However, the continuity of $f(x)$ at $x = a$ requires a further condition that the value of $f(x)$ just at $x = a$ is to be defined. In symbol, we can say that

$$f(x) \text{ jas a limit at } x = a \Rightarrow \lim_{x \to a-0} f(x) = \lim_{x \to a+0} f(x),$$

and

$$f(x) \text{ is continuous at } x = a \Rightarrow \lim_{x \to a-0} f(x) = \lim_{x \to a+0} f(x) = f(a).$$

Given a function $f(x)$ on a domain D, the limit of $f(x)$ may and may not lie in D. In contrast, the continuity of $f(x)$ should be defined only at points that must lie in D. An illustrative example is given below.

Example:
Suppose a function given by

$$f(x) = x \text{ for all but } x = 1.$$

It has a limit at $x = 1$ as

$$\lim_{x\to\infty} f(x) = 1.$$

But there is no way to examine its continuity at $x = 1$, because $x = 1$ is out of the defining domain. ■

When $f(x)$ is continuous, we can say that $f(x)$ belongs to the class of functions designated by the symbol C. Then, it follows that

$$f(x) \in C \text{ at } x = a \iff \lim_{x\to a} f(x) = f(a).$$

If the symbol $x \to a$ appearing in the right statement is replaced by $x \to a+$ (or $x \to a-$), $f(x)$ is said to be **continuous on the right** (or **left**) at $x = a$. We encounter the latter kind of a limit, particularly when considering the continuity of a function defined within a finite interval $[a,b]$; we say that

$$f(x) \in C \text{ on } [a,b] \iff f(x) \in C \text{ on } (a,b), \ \lim_{x\to a+} f(x) = f(a), \text{ and } \lim_{x\to b-} f(x) = f(b).$$

We also say that a function $f(x)$ on $[a,b]$ is **piecewise continuous** if the following two conditions are satisfied.

1. $f(x)$ is continuous on $[a,b]$ except at a finite number of points $x_1, x_2, \cdots, x_n$.
2. At each of the points $x_1, x_2, \cdots, x_n$, there exist both the left-hand and right-hand limits of $f(x)$ defined by

$$f(x_k - 0) = \lim_{x=x_k-0} f(x), \quad f(x_k + 0) = \lim_{x=x_k+0} f(x).$$

D.3 Derivative of function

Here is given a rigorous definition of the derivative of a real function.

Definition (Derivative of a function):
If the limit

$$\lim_{x\to a} \frac{f(x) - f(a)}{x - a}$$

exists, it is called the **derivative** of $f(x)$ at $x = a$, and denoted by $f'(a)$. ■

The function $f(x)$ is said to be **differentiable** at $x = a$ if $f'(a)$ exists.

Similar to the case of one-sided limits, it is possible to define **one-sided derivative** of real functions such as

$$\begin{aligned} f'(a+) &= \lim_{x\to a+} \frac{f(x)-f(a)}{x-a}, \\ f'(a-) &= \lim_{x\to a-} \frac{f(x)-f(a)}{x-a}. \end{aligned}$$

Theorem:
If $f(x)$ is differentiable at $x = a$, then it is continuous at $x = a$ (Note that the converse is not true).[1] ■

Proof:
Assume $x \neq a$. Then

$$f(x) - f(a) = \frac{f(x)-f(a)}{x-a}(x-a).$$

From hypothesis, each function $[f(x) - f(a)]/(x-a)$ as well as $x-a$ has the limit at $x = a$. Hence we obtain

$$\lim_{x\to a}[f(x)-f(a)] = \lim_{x\to a}\frac{f(x)-f(a)}{x-a}\cdot\lim_{x\to a}(x-a) = f'(a)\times 0 = 0.$$

This consequences that

$$\lim_{x\to a} f(x) = f(a),$$

that is, $f(x)$ is continuous at $x = a$. ■

We mean by C^n functions that all derivatives of order $\leq n$ exist; this is denoted by

$$f(x) \in C^n \quad \Longleftrightarrow \quad f^{(n)}(x) \in C.$$

Such $f(x)$ is said C^n function or of class C^n.

Example 1:

$$f(x) = \begin{cases} 0, & x < 0 \\ x, & x \geq 0 \end{cases} \quad \Rightarrow \quad f(x) \in C^0(=C), \text{ but } f(x) \notin C^1 \text{ at } x = 0. \quad ■$$

Example 2:

$$f(x) = \begin{cases} 0, & x < 0 \\ x^2, & x \geq 0 \end{cases} \quad \Rightarrow \quad f(x) \in C^1, \text{ but } f(x) \notin C^2 \text{ at } x = 0. \quad ■$$

Example 3:
Taylor series expansion of functions $f \in C^n$ is given by

$$f(x) = \sum_{k\leq n} \frac{1}{k!}\left.\frac{\partial f}{\partial x^k}\right|_{x=x_0}(x-x_0)^k + o\left(|x-x_0|^n\right). \quad ■$$

---

[1]That the converse is false can be seen by considering $f(x) = |x|$; it is continuous at $x = 0$ but not differentiable.

D.4 Smooth function

We now further introduce a new class of functions, for which the derivative is continuous over the defining domain.

Definition (Smooth function):
The function $f(x)$ is said to be **smooth** for any $x \in [a,b]$ if $f'(x)$ exists and is continuous on $[a,b]$. ∎

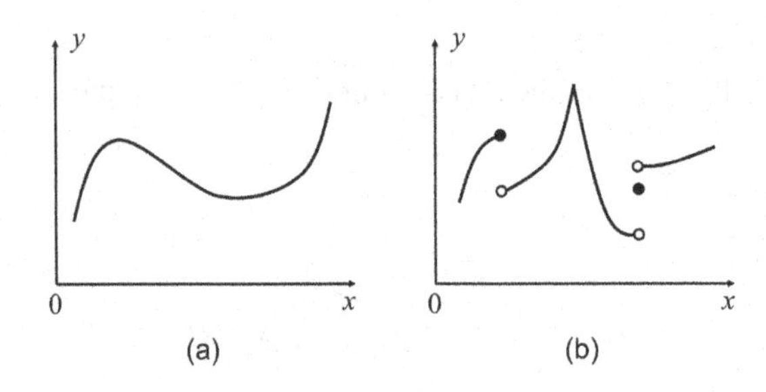

Figure D.2: (a) A continuous function $y = f(x)$. (b) A piecewise smooth function $y = f(x)$ having two discontinuous points and one corner.

In geometrical language, the smoothness of a function means that the direction of the tangent changes continuously, without jumps, as it moves along the curve $y = f(x)$. (See Fig. D.2.) Thus, the graph of a smooth function is a smooth curve without any point at which the curve has two distinct tangents.

Similar to the case of piecewise continuity, the function $f(x)$ is said to be **piecewise smooth** on the interval $[a,b]$ if $f(x)$ and $f'(x)$ are both piecewise continuous on $[a,b]$. The graph of a piecewise smooth function can have a finite number of points (so-called the **corners**) at which the derivative show jumps. (See Fig. D.2.)

Appendix E

Function Sequence

E.1 Pointwise convergence

In Appendix E, we focus on convergence properties of sequences $\{f_n\}$ consisting of real-valued *functions* $f_n(x)$ of a real variable x. Suppose that, for each $n \in \mathbb{N}$, we have a function $f_n(x)$ defined on a domain $D \subseteq \mathbb{R}$. If the sequence $\{f_n\}$ converges for every $x \in D$, the sequence of functions is said to **converges pointwise** on D, and the function defined by

$$f(x) = \lim_{n\to\infty} f_n(x) \tag{E.1}$$

is called the **pointwise limit** of $\{f_n\}$. The formal definition is given below.

Definition (Pointwise convergence):
The sequence of functions $\{f_n\}$ is said to **converge pointwise** to f on D if, given $\varepsilon > 0$, there is a natural number $N = N(\varepsilon, x)$ (whose value depends on ε and x) such that

$$n > N \;\Rightarrow\; |f_n(x) - f(x)| < \varepsilon. \quad \blacksquare \tag{E.2}$$

Example:
Suppose a sequence $\{f_n\}$ consisting of the function

$$f_n(x) = x^n, \tag{E.3}$$

which is defined on a closed interval $[0,1]$. The sequence converges pointwise to

$$f(x) = \lim_{n\to\infty} f_n(x) = \begin{cases} 0 & \text{for } 0 \le x < 1, \\ 1 & \text{at } x = 1. \end{cases} \tag{E.4}$$

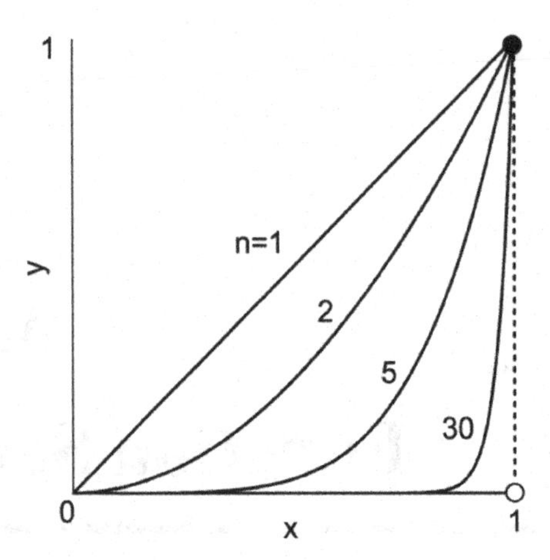

Figure E.1: Converging behavior of $f_n(x) = x^n$ given in (E.3).

Figure E.1 illustrates the converging behavior of $f_n(x)$ with increasing n. ■

The important fact is that under pointwise convergence, continuity of functions of $f_n(x)$ is not preserved at the limit of $n \to \infty$. In fact, $f_n(x)$ given in (E.3) is continuous for each n over the whole interval $[0,1]$, whereas the limit $f(x)$ given in (E.4) is discontinuous at $x = 1$. This indicates that interchanging the order of the limiting processes under pointwise convergence may produce different results as expressed by

$$\lim_{x\to 1}\lim_{n\to\infty} f_n(x) \neq \lim_{n\to\infty}\lim_{x\to 1} f_n(x). \tag{E.5}$$

Similar phenomena possibly occur as to integrability and differentiability of terms of functions $f_n(x)$. That is, under pointwise convergence, the limit of a sequence of integrable or differentiable functions may not be integrable or differentiable, respectively. Illustrative examples are given below.

Example 1:
Consider the function:

$$f_n(x) = nx(1-x^2)^n, \quad x \in [0,1]. \tag{E.6}$$

We show below that exchanging the order of the limiting process $n \to \infty$ and integration gives different results.

The given function is integrable for each n so that

$$\begin{aligned}\int_0^1 f_n(x)dx &= n\int_0^1 x(1-x^2)^n dx = \left[\frac{-n}{2(n+1)}(1-x^2)^{n+1}\right]_0^1 \\ &= \frac{n}{2(n+1)} \to \frac{1}{2}.\end{aligned}$$

Contrary, the limit given by

$$f(x) = \lim_{n\to\infty} f_n(x) = 0 \quad \text{for all } x \in [0,1] \tag{E.7}$$

yields $\int_0^1 f(x)dx = 0$. It is thus concluded that

$$\lim_{n\to\infty} \int_0^1 f_n(x)dx \neq \int_0^1 \lim_{n\to\infty} f_n(x)dx; \tag{E.8}$$

that is, interchanging the order of integration and limiting processes is not allowed in general under pointwise convergence. ■

Example 2:
For $f_n(x)$ given by

$$f_n(x) = \begin{cases} -1 & x < -\frac{1}{n}, \\ \sin\left(\frac{n\pi x}{2}\right) & -\frac{1}{n} < x < \frac{1}{n}, \\ 1 & x > \frac{1}{n}, \end{cases} \tag{E.9}$$

we now check the continuity of its limit $f(x) = \lim_{n\to\infty} f_n(x)$ at $x = 0$.

$f_n(x)$ is differentiable for any $x \in \mathbb{R}$ for all n, and thus is continuous at $x = 0$ for all n. However, its limit

$$f(x) = \begin{cases} -1 & x < 0, \\ 0 & x = 0, \\ 1 & x > 0, \end{cases} \tag{E.10}$$

is not continuous at $x = 0$. Hence, as for the sequence of functions $\{f_n(x)\}$, the order of the limiting process $n \to \infty$ and the differentiation with respect to x is not exchangeable. ■

E.2 Uniform convergence

We have known that if the sequence $\{f_n\}$ is pointwise convergent to $f(x)$ on $x \in D$, it is possible to choose $N(x)$ for any small ε such that

$$m > N(x) \;\Rightarrow\; |f_m(x) - f(x)| < \varepsilon. \tag{E.11}$$

In general, the least value of $N(x)$ that satisfies (E.11) will depend on x. But in certain cases, we can choose N *independently* of x such that $|f_m(x) - f(x)| < \varepsilon$ for all $m > N$ and *for all* x over the domain D. If this is true for any small ε, the sequence $\{f_n\}$ is said to **uniformly converge** to $f(x)$ on D. The formal definition is given below.

Definition (Uniform convergence):
The sequence $\{f_n\}$ of real functions on $D \subseteq \mathbb{R}$ **converges uniformly** to a function f on D if, given $\varepsilon > 0$, there is a positive integer $N = N(\varepsilon)$ (which depends on ε) such that

$$n > N \Rightarrow |f_n(x) - f(x)| < \varepsilon \text{ for all } x \in D. \quad \blacksquare \tag{E.12}$$

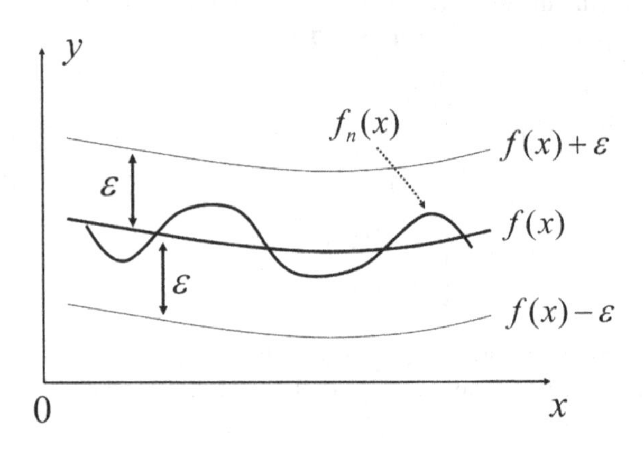

Figure E.2: A function $y = f_n(x)$ contained overall within a band of width less than 2ε.

Emphasis is put on the fact that the integer $N = N(\varepsilon, x)$ in the pointwise convergence *depends* on x in general, while that $N = N(\varepsilon)$ in the uniform one is *independent* of x. Under uniform convergence, therefore, the graph of $y = f_n(x)$ is embedded into a band of width less than 2ε centered about the graph of $y = f(x)$ over the whole domain D, by taking n large enough (See Fig. E.2).

Keypoint:
1. Pointwise convergence: $N = N(x, \varepsilon)$.
2. Uniform convergence: $N = N(\varepsilon)$ independent of x.

The definition of uniform convergence noted above is equivalent to the following statement.

Theorem:
The sequence $\{f_n\}$ of real functions on $D \subseteq \mathbb{R}$ converges uniformly to f on D if and only if

$$\sup_{x \in D} |f_n(x) - f(x)| = 0 \text{ as } n \to \infty. \quad \blacksquare \tag{E.13}$$

E.3 Cauchy criterion for function series

As in the case of real sequences, Cauchy criterion is available for testing uniform convergence for sequences of functions.

Theorem (Cauchy criterion for uniform convergence):
The sequence of f_n defined on $D \subseteq \mathbb{R}$ converges uniformly to f on D if and only if, given $\varepsilon > 0$, there is a positive integer $N = N(\varepsilon)$ such that

$$m, n > N \;\Rightarrow\; |f_m(x) - f_n(x)| < \varepsilon \text{ for all } x \in D, \tag{E.14}$$

or equivalently,

$$m, n > N \;\Rightarrow\; \sup_{x \in D} |f_m(x) - f_n(x)| < \varepsilon. \quad \blacksquare \tag{E.15}$$

Proof:
Suppose that $f_n(x)$ converges uniformly to $f(x)$ on D. Let $\varepsilon > 0$ and choose $N \in \mathbf{N}$ such that

$$n > N \;\Rightarrow\; |f_n(x) - f(x)| < \frac{\varepsilon}{2} \text{ for all } x \in D. \tag{E.16}$$

If $m, n \geq N$, we have

$$|f_n(x) - f_m(x)| \leq |f_n(x) - f(x)| + |f(x) - f_n(x)| < \varepsilon \text{ for all } x \in D. \tag{E.17}$$

This result implies that if $f_n(x)$ is uniformly convergent to $f(x)$ on D, there exists an N satisfying (E.14) for any small ε.

Next we consider the converse. Suppose that $\{f_n\}$ satisfies the criterion given by (E.14). Then, for each point of $x \in D$ $\{f_n\}$ forms a Cauchy sequence, and thus converges pointwise to

$$f(x) = \lim_{n \to \infty} f_n(x) \text{ for all } x \in D. \tag{E.18}$$

We shall show that this convergence is uniform. Let $n > N$ be fixed and take the limit $m \to \infty$ in (E.14) to obtain

$$n > N \;\Rightarrow\; |f_n(x) - f(x)| < \varepsilon \text{ for all } x \in D, \tag{E.19}$$

where N is independent of x. Therefore, we conclude that the convergence of $\{f_n\}$ to f is uniform. $\blacksquare$

Example:
We show below that the sequence of functions $\{f_n\}$ defined by

$$f_n(x) = nxe^{-nx}, \tag{E.20}$$

converges uniformly to $f(x) = 0$ on $x > 0$. In view of the previous theorem, let us first consider the validity of the relation

$$\sup\{f_n(x) : \; x \geq a\} = 0 \text{ as } n \to \infty, \tag{E.21}$$

where $a > 0$. To prove it, we consider the derivative

$$f_n{}'(x) = ne^{-nx}(1 - nx). \tag{E.22}$$

It follows from (E.22) that $x = 1/n$ is the only critical point of f_n. Now we choose a positive integer N such that $a > 1/N$. Then, the function f_n for each $n \geq N$ has no critical point on $x \geq a$, and is monotonically decreasing. Therefore, the maximum of $f_n(x)$ is attained at $x = a$ for any $n > N$, which consequences that

$$\sup_{x \in [a,\infty)} f_n(x) = f_n(a) = nae^{-na} \to 0. \quad (n \to \infty) \tag{E.23}$$

This holds for any $a > 0$; hence we conclude that f_n converges uniformly to 0 on $(0, \infty)$, i.e., on $x > 0$. ■

Notice that the range of uniform convergence of (E.20) is the open interval $(0, \infty)$, not the closed one $[0, \infty)$. Since in the latter case, we have

$$\sup_{x \in [0,\infty)} f_n(x) = f_n\left(\frac{1}{n}\right) = \frac{1}{e} \not\to 0, \tag{E.24}$$

which clearly means that $\{f_n\}$ does not converge uniformly on $[0, \infty)$.

Appendix F

Uniformly Convergent Sequence of Functions

F.1 Continuity of the limit function

The most important feature of **uniform convergence** is that it overcomes some shortcomings of **pointwise convergence** demonstrated in Section E.1. Namely, pointwise convergence do neither preserve continuity, integrability, nor differentiability of terms of functions $f_n(x)$. Contrariwise, the manner of convergence is *uniform*, then the continuity, integrability, and differentiability of $f_n(x)$ are preserved at $n \to \infty$ under loose conditions. We now examine the situation under uniform convergence, starting at the continuity of $f_n(x)$.

Theorem:
Suppose that f_n converges uniformly to f on $D \subseteq \mathbb{R}$. If f_n is continuous at $c \in D$, then f is also continuous at c. ■

Note that the uniform convergence of f_n on D is only a sufficient condition, not necessary, for f to be continuous. In fact, if f_n is not uniformly convergent on D, then its limit f may and may not continuous at $c \in D$.

The contraposition of the theorem also warrants attention. It tells us that, if the limit function f is discontinuous, the convergence of f_n is not uniform. Example in §E.1 has demonstrated such a sequence.

To prove the continuity of $f(x)$ at $x = c$, it suffices to see that

$$\lim_{x \to c} f(x) = \lim_{x \to c} \lim_{n \to \infty} f_n(x) = \lim_{n \to \infty} \lim_{x \to c} f_n(x) = \lim_{n \to \infty} f_n(c) = f(c). \tag{F.1}$$

In (F.1), we have used the exchangeability of limiting processes expressed by

$$\lim_{x \to c} \lim_{n \to \infty} f_n(x) = \lim_{n \to \infty} \lim_{x \to c} f_n(x), \tag{F.2}$$

which follows from the Lemma below.

Lemma:
Let c be an accumulation point of $D \subseteq \mathbb{R}$ and suppose that f_n converges uniformly to f on $D\backslash\{c\}$. If

$$\lim_{x\to c} f_n(x) = \ell_n \tag{F.3}$$

exists for each n, then

1. (ℓ_n) is convergent, and
2. $\lim_{x\to c} f(x)$ exists and coincides with $\lim_{n\to\infty} \ell_n$, i.e.,

$$\lim_{n\to\infty}\lim_{x\to c} f_n(x) = \lim_{x\to c}\lim_{n\to\infty} f_n(x). \quad \blacksquare \tag{F.4}$$

Proof (of Lemma):
Let $\varepsilon > 0$. Since $\{f_n\}$ converges uniformly on $D\backslash\{c\}$, it satisfies the Cauchy criterion; that is, there is a positive integer N such that

$$m,n > N \ \Rightarrow \ |f_n(x) - f_m(x)| < \varepsilon \ \text{ for all } x \in D\backslash\{c\}. \tag{F.5}$$

Take the limit $x \to c$ in (F.5) to obtain

$$m,n > N \ \Rightarrow \ |\ell_n - \ell_m| < \varepsilon. \tag{F.6}$$

This implies that (ℓ_n) is a Cauchy sequence and thus being convergent. This result proves the statement (i) above.

To prove (ii), let

$$\ell = \lim_{n\to\infty} \ell_n. \tag{F.7}$$

Set $n = N$ and $m \to \infty$ in (F.3), (F.5) and (F.6), to have the results:

$$\lim_{x\to c} f_N(x) = \ell_N, \tag{F.8}$$

$$|f_N(x) - f(x)| < \varepsilon \ \text{ for all } x \in D\backslash\{c\}, \tag{F.9}$$

and

$$|\ell_N - \ell| < \varepsilon. \tag{F.10}$$

The existence of (F.8) implies that there exists a $\delta > 0$ such that

$$|x - c| < \delta \text{ with } x \in D\backslash\{c\} \ \Rightarrow \ |f_N(x) - \ell_N| < \varepsilon. \tag{F.11}$$

Using (F.9), (F.10) and (F.11), we obtain

$$\begin{aligned} & |x-c| < \delta \ \text{ with } x \in D\backslash\{c\} \\ \Rightarrow \quad & |f(x) - \ell| \le |f(x) - f_N(x)| + |f_N(x) - \ell_N| + |\ell_N - \ell| < 3\varepsilon. \end{aligned}$$

The result gives a consequence that

$$\lim_{x\to c} f(x) = \ell, \tag{F.12}$$

which is equivalent to the desired result of (F.4). $\blacksquare$

F.2 Integrability of the limit function

We have known that the limit function $f(x)$ becomes continuous if the sequence $(f_n(x))$ of continuous functions is uniformly convergent. This immediately consequences the following theorem.

Theorem:
Suppose f_n be integrable on $[a,b]$ for each n. Then,[1] if f_n converges uniformly to f on $[a,b]$, the limit function f is also integrable so that

$$\int_a^b f(x)dx = \int_a^b \lim_{n\to\infty} f_n(x)dx = \lim_{n\to\infty} \int_a^b f_n(x)dx. \quad \blacksquare \tag{F.13}$$

Proof:
Since f_n for each n is integrable on $[a,b]$, it is continuous (piecewise, at least) on $[a,b]$. Thereby $f(x)$ is also continuous (piecewise at least) on $[a,b]$ in view of the theorem given in §F.1. And thus $f(x)$ is integrable on $[a,b]$. Furthermore, we observe that

$$\begin{aligned} \left| \int_a^b f_n(x)dx - \int_a^b f(x)dx \right| &\le \int_a^b |f_n(x) - f(x)|\, dx \\ &\le \int_a^b \sup_{x\in[a,b]} |f_n(x) - f(x)|\, dx \\ &\le (b-a) \sup_{x\in[a,b]} |f_n(x) - f(x)|. \end{aligned}$$

The uniform convergence of $\{f_n\}$ ensures that

$$\sup_{x\in[a,b]} |f_n(x) - f(x)| \to 0 \text{ as } n \to \infty, \tag{F.14}$$

which immediately consequences the desired result of (F.13). $\blacksquare$

Note again that uniform convergence is a *sufficient* condition, not *necessary*, for the equation (F.13) to be valid. Therefore, (F.13) may be valid even in the absence of uniform convergence. For instance, the convergence of $\{f_n\}$ with $f_n(x) = x^n$ on $[0,1]$ is not uniform but we have

$$\int_0^1 f_n(x)dx = \int_0^1 x^n dx = \frac{1}{n+1} \to 0 = \int_0^1 f(x)dx. \tag{F.15}$$

F.3 Differentiability of the limit function

Followed by the two previous subsections, the reader may expect results for differentiability as similar to continuity and integrability, i.e., they may be tempted

[1] The conditions on f_n stated in the theorem will be significantly relaxed when we take up the Lebesgue integral as accounted for in Section 6.4.2.

to say that the differentiability of terms of functions $f_n(x)$ would be preserved if $\{f_n\}$ converges uniformly to f. However, this is not the case. In fact, even if f_n converges uniformly to f on $[a,b]$ and f_n is differentiable at $c \in [a,b]$, it may occur that

$$\lim_{n\to\infty} f_n'(c) \neq f'(c). \tag{F.16}$$

Consider the following example.

Example:
Suppose the sequence $\{f_n\}$ defined by

$$f_n(x) = \sqrt{x^2 + \frac{1}{n^2}}, \quad x \in [-1,1]. \tag{F.17}$$

Clearly (F.17) is differentiable for each n, and the sequence $\{f_n\}$ converges uniformly on $[-1,1]$ to

$$f(x) = |x|, \tag{F.18}$$

since

$$\begin{aligned} |f_n(x) - f(x)| &= \sqrt{x^2 + \frac{1}{n^2}} - \sqrt{x^2} \\ &= \frac{\frac{1}{n^2}}{\sqrt{x^2 + \frac{1}{n^2}} + \sqrt{x^2}} \leq \frac{1}{n} \to 0, \quad \text{for all } x \in [-1,1]. \end{aligned}$$

However, the limit function f of (F.18) is not differentiable at $x = 0$. Hence, the desired result

$$\lim_{n\to\infty} f_n'(x) = f'(x) \tag{F.19}$$

breaks down at $x = 0$. ■

The following theorem gives sufficient conditions for (F.19) to be satisfied. Emphasis should be placed on that it requires the uniform convergence of the *derivatives* f_n', not of the functions f_n themselves.

Theorem:
Suppose $\{f_n\}$ be a sequence of differentiable functions on $[a,b]$ which converges at a certain point $x_0 \in [a,b]$. If the sequence $\{f_n'\}$ is uniformly convergent on $[a,b]$, then

1. $\{f_n\}$ is also uniformly convergent on $[a,b]$ to f,
2. f is differentiable on $[a,b]$, and
3. $\lim_{n\to\infty} f_n'(x) = f'(x)$. ■

Again we should pay attention to that the uniform convergence of $\{f_n'\}$ is just a sufficient, not necessary, condition. This fact is seen by considering the sequence

$$f_n(x) = \frac{x^{n+1}}{n+1} \quad x \in (0,1). \tag{F.20}$$

This converges uniformly to 0, and its derivative $f_n'(x) = x^n$ converges also to 0. Hence, the conclusions **i)-iii)** given in the theorem above are all satisfied. However, the convergence of $\{f_n'\}$ is not uniform.

Proof:
Let $\varepsilon > 0$. From the convergence of $\{f_n(x_0)\}$ and the uniform convergence of $\{f_n'\}$, we deduce that there is an $N \in \mathbb{N}$ such that

$$m,n > N \;\Rightarrow\; |f_n'(x) - f_m'(x)| < \varepsilon \text{ for all } x \in [a,b], \tag{F.21}$$

and

$$m,n > N \;\Rightarrow\; |f_n(x_0) - f_m(x_0)| < \varepsilon. \tag{F.22}$$

Given any two points $x,t \in [a,b]$, it follows from the **mean value theorem**, applied to $f_n - f_m$, that there is a point c between x and t such that

$$f_n(x) - f_m(x) - [f_n(t) - f_m(t)] = (x-t)\left[f_n'(c) - f_m'(c)\right]. \tag{F.23}$$

Using (F.21), we have

$$m,n > N \;\Rightarrow\; |f_n(x) - f_m(x) - [f_n(t) - f_m(t)]| < \varepsilon |x-t|, \tag{F.24}$$

From (F.22) and (F.24), it follows that

$$\begin{aligned} |f_n(x) - f_m(x)| &\leq |f_n(x) - f_m(x) - [f_n(x_0) - f_m(x_0)]| + |f_n(x_0) - f_m(x_0)| \\ &< \varepsilon |x - x_0| + \varepsilon \\ &< \varepsilon(b-a+1) = C\varepsilon, \quad \text{for all } x \in [a,b]. \end{aligned}$$

This means that the sequence $\{f_n\}$ converges uniformly to some limit f. Hence the statement of (i) has been proven.

Next we work on the proofs of (ii) and (iii). For any fixed point $x \in [a,b]$, define

$$\begin{aligned} f_n(t) &= \frac{f_n(t) - f_n(x)}{t-x}, \quad t \in [a,b]/\{x\}, \text{ and} \\ g(t) &= \frac{f(t) - f(x)}{t-x}, \quad t \in [a,b]/\{x\}. \end{aligned}$$

Clearly, $f_n \to g$ as $n \to \infty$; furthermore, if $m,n \geq N$, the result of (F.24) tells us that

$$|f_n(t) - f_m(t)| < \varepsilon \text{ for all } t \in [a,b]/\{x\}. \tag{F.25}$$

In view of Cauchy criterion, therefore, we see that f_n converges uniformly to g on $[a,b]\backslash\{x\}$. Now we observe that

$$\lim_{t\to x} f_n(t) = f_n'(x) \text{ for all } n \in \mathbb{N}. \tag{F.26}$$

Then, uniform convergence of f_n assures to take the limit of $n \to \infty$ in (F.26) followed by exchanging the order of the limit processes, which yields

$$\lim_{n\to\infty}\lim_{t\to x} f_n(t) = \lim_{t\to x} g(t) = \lim_{t\to x}\frac{f(t)-f(x)}{t-x} = f'(x) = \lim_{n\to\infty} f_n'(x). \tag{F.27}$$

This proves that f is differentiable at x, and also that

$$f'(x) = \lim_{n\to\infty} f_n'(x). \quad \blacksquare \tag{F.28}$$

Appendix G

Function Series

G.1 Infinite series of functions

We close the train of Appendixes by considering convergence properties of series of real-valued functions. Suppose a sequence $\{f_n\}$ of functions defined on $D \subseteq \mathbb{R}$. In analogy with series of real numbers, we can define a series of functions by

$$S_n(x) = \sum_{k=1}^{n} f_k(x), \quad x \in D, \tag{G.1}$$

which gives a sequence $\{S_n\} = \{S_1, S_2, \cdots\}$.

As n increases, the sequence $\{S_n\}$ may and may not converge to a finite value, depending on the feature of functions $f_k(x)$ as well as the point x in question. If the sequence converges for each point $x \in D$ (i.e., converges pointwise on D), then the limit of S_n is called the **sum of the infinite series of functions** $f_k(x)$, and is denoted by

$$S(x) = \lim_{n\to\infty} S_n(x) = \sum_{k=1}^{\infty} f_k(x), \quad x \in D. \tag{G.2}$$

It is obvious that the convergence of the series $S_n(x)$ implies the pointwise convergence $\lim_{n\to\infty} f_n(x) = 0$ on D. A series $\{S_n\}$ which does not converges at a point $x \in D$ is said to diverge at that point.

Applied to series of functions, Cauchy criterion for uniform convergence takes the following form:

Theorem (Cauchy criterion for series of functions):
The series S_n is uniformly convergent on D if and only if, for every small $\varepsilon > 0$, there

is a positive integer N such that

$$\begin{aligned} & n > m > N \\ \Rightarrow \quad & |S_n(x) - S_m(x)| = \left| \sum_{k=m+1}^{n} f_k(x) \right| < \varepsilon \quad \text{for all } x \in D. \quad \blacksquare \end{aligned}$$

Set $n = m+1$ in the above criterion to obtain

$$n > N \;\Rightarrow\; |f_n(x)| < \varepsilon \text{ for all } x \in D. \tag{G.3}$$

This results implies that the uniform convergence of $f_n(x) \to 0$ on D is a necessary condition for the convergence of $S_n(x)$ to be uniform on D.

G.2 Properties of uniformly convergent series of functions

When a given series of functions $\sum f_k(x)$ enjoys uniform convergence, the properties of the sum $S(x)$ as to continuity, integrability, and differentiability can be easily inferred from the properties of the separate terms $f_k(x)$. In fact, applying Theorems given in §F.1-F.3 to the sequence $\{S_n\}$, and using the linearity regarding the limiting process, integration, and differentiation, we obtain the following parallel theorems:

Theorem (Continuity of a sum of functions):
Suppose $f_k(x)$ be continuous for each k. If the sequence $\{S_n\}$ of the series

$$S_n(x) = \sum_{k=1}^{n} f_k(x) \tag{G.4}$$

converges uniformly to $S(x)$, then $S(x)$ is also continuous so that

$$\lim_{t \to x} S(t) = \lim_{t \to x} \sum_{k=1}^{\infty} f_k(t) = \sum_{k=1}^{\infty} \lim_{t \to x} f_k(t). \quad \blacksquare \tag{G.5}$$

Theorem (Integrability of a sum of functions):
Suppose f_k to be integrable on $[a,b]$ for all k. If $\{S_n\}$ converges uniformly to S on $[a,b]$, then we have

$$\int_a^b S(x)dx = \int_a^b \sum_{k=1}^{\infty} f_k(x)dx = \sum_{k=1}^{\infty} \int_a^b f_k(x)dx. \quad \blacksquare \tag{G.6}$$

Theorem (Differentiability of a sum of functions):
Let f_k be differentiable on $[a,b]$ for each k, and suppose that $\{S_n\}$ converges to S at some point $x_0 \in [a,b]$. If the series $\sum f_k'$ is uniformly convergent on $[a,b]$, then $S_n(x)$ is also uniformly convergent on $[a,b]$, and the sum $S(x)$ is differentiable on $[a,b]$ so that

$$\frac{d}{dx} S(x) = \frac{d}{dx}\left[\sum_{k=1}^{\infty} f_k(x)\right] = \sum_{k=1}^{\infty} \frac{df_k(x)}{dx} \quad \text{for all } x \in [a,b]. \quad \blacksquare \tag{G.7}$$

Observe that the second and third theorems provide a sufficient condition for performing term-by-term integration and differentiation, respectively, of infinite series of functions. Without uniform convergence, such term-by-term calculations do not work at all.

Appendix H

Matrix Eigenvalue Problem

H.1 Eigenvalue and eigenvector

As a pedagogical preparation, we summarize below the essence of eigenvalue analysis for finite dimensional matrices. The accumulated understanding on the finite dimensional case will be of help in developing the spectrum theory of continuous linear operators. Elements in the n-dimensional number space $\mathbb{C}^n$ are written by

$$\boldsymbol{u} = \begin{bmatrix} u_1 \\ \vdots \\ u_n \end{bmatrix} \in \mathbb{C}^n, \quad \boldsymbol{v} = \begin{bmatrix} v_1 \\ \vdots \\ v_n \end{bmatrix} \in \mathbb{C}^n, \tag{H.1}$$

and the inner product of the two vectors reads

$$(\boldsymbol{u}, \boldsymbol{v}) = \sum_{j=1}^{n} u_j \overline{v}_j, \tag{H.2}$$

where $\overline{v}_j$ is complex conjugate of v_j. In addition, the norm in $\mathbb{C}^n$ is defined by

$$\|\boldsymbol{u}\| = \sqrt{(\boldsymbol{u}, \boldsymbol{u})}. = \sqrt{\sum_{j=1}^{n} |u_j|^2}. \tag{H.3}$$

Under the norm, the space of $\mathbb{C}^n$ serves as a Hilbert space owing to the completeness of $\mathbb{C}^n$. Let us consider a linear operator $\boldsymbol{A}$ in $\mathbb{C}^n$, which associate every $\boldsymbol{u} \in \mathbb{C}^n$ with $\boldsymbol{A}\boldsymbol{u} \in \mathbb{C}^n$. Using the matrix representation, the mapping from $\boldsymbol{u}$ to $\boldsymbol{A}\boldsymbol{u}$ is written by

$$\boldsymbol{A}\boldsymbol{u} = \begin{bmatrix} a_{11} & \cdots & a_{1n} \\ \vdots & \ddots & \vdots \\ a_{n1} & \cdots & a_{nn} \end{bmatrix} \boldsymbol{u}. \tag{H.4}$$

What does it mean that a matrix $\boldsymbol{A}$ acting on $\mathbb{C}^n$ has an eigenvalue $\lambda \in \mathbb{C}$? It indicates the existence of a non-zero element $bmx \in \mathbb{C}^n$ that satisfies the proportional relation

$$\boldsymbol{A}\boldsymbol{x} = \lambda \boldsymbol{x}. \tag{H.5}$$

Such the element $\boldsymbol{x} \in \mathbb{C}^n$ is called the eigenvector of the matrix $\boldsymbol{A}$ with respect to the eigenvalue λ. If it is possible to find a set of eigenvectors that are orthonormal to each other, the set of eigenvectors provides an efficient tool for solving the eigenvalue problem of the matrix $\boldsymbol{A}$. Actually, it is an **orthonormal basis** of the matrix $\boldsymbol{A}$, enabling us to solve a generalized eigenvalue problem such as

$$\boldsymbol{A}\boldsymbol{u} = \boldsymbol{v}. \tag{H.6}$$

H.2 Hermite matrix

Among many classes of matrices having eigenvalues, **Hermite matrices** (or often called **self-adjoint matrices**) will be focused on in the meantime. This is because in the following argument, we would like to generalize the concept of Hermite matrices to that applicable to infinite dimensional Hilbert spaces.

The notion of Hermite matrices is introduced on the basis of the **adjoint matrix**. Given an n-by-n matrix $\boldsymbol{A}$ having complex-valued entries,

$$\boldsymbol{A} = \begin{bmatrix} a_{11} & \cdots & a_{1n} \\ \vdots & \ddots & \vdots \\ a_{n1} & \cdots & a_{nn} \end{bmatrix} \tag{H.7}$$

its adjoint matrix is defined by

$$\boldsymbol{A}^* = \begin{bmatrix} \overline{a_{11}} & \cdots & \overline{a_{1n}} \\ \vdots & \ddots & \vdots \\ \overline{a_{n1}} & \cdots & \overline{a_{nn}} \end{bmatrix}. \tag{H.8}$$

Observe that all the entries in $\boldsymbol{A}^*$ are complex conjugate to those in $\boldsymbol{A}$. In general, the adjoint matrix $\boldsymbol{A}^*$ of a matrix $\boldsymbol{A}$ satisfies the relation

$$(\boldsymbol{A}\boldsymbol{u}, \boldsymbol{v}) = (\boldsymbol{u}, \boldsymbol{A}^*\boldsymbol{v}). \tag{H.9}$$

We are ready to give a precise definition of Hermite matrices.

Definition (Hermite matrix):
An **Hermite matrix** $\boldsymbol{A}$ is defined as those satisfying the condition of $\boldsymbol{A} = \boldsymbol{A}^*$. ■

Every Hermite matrix has a set of eigenvalues. The primary feature of those eigenvalues is written by the following statement.

Theorem:
Eigenvalues of an Hermite matrix are all real valued. ■

It is known that in elementary linear algebra, every Hermite matrix $\boldsymbol{A}$ can be diagonalized through an appropriate **unitary matrix**[1] $\boldsymbol{U}$ by

$$\boldsymbol{U}^*\boldsymbol{A}\boldsymbol{U} = \begin{bmatrix} \lambda_1 & 0 & \cdots & 0 \\ 0 & \ddots & \ddots & \vdots \\ \vdots & \ddots & \ddots & 0 \\ 0 & \cdots & 0 & \lambda_n \end{bmatrix}, \tag{H.10}$$

where the real numbers $\lambda_1, \cdots, \lambda_n$ are eigenvalues of $\boldsymbol{A}$. Suppose that the unitary matrix $\boldsymbol{U}$ given above is represented by

$$\boldsymbol{U} = \begin{bmatrix} u_{11} & \cdots & u_{1n} \\ \vdots & \ddots & \vdots \\ u_{n1} & \cdots & u_{nn} \end{bmatrix} = \left[\boldsymbol{u}^{(1)}, \cdots, \boldsymbol{u}^{(n)}\right], \quad \boldsymbol{u}^{(j)} = \begin{bmatrix} u_{1j} \\ \vdots \\ u_{nj} \end{bmatrix}. \tag{H.11}$$

It can be proved that the set of vectors, $\left\{\boldsymbol{u}^{(1)}, \cdots, \boldsymbol{u}^{(n)}\right\}$, is an **orthonormal basis** of the n-dimensional space. Once obtaining the orthonormal basis, it is possible to decompose an arbitrary vector $\boldsymbol{x}$ in terms of the basis as follows.

$$\boldsymbol{x} = \left(\boldsymbol{x}, \boldsymbol{u}^{(1)}\right)\boldsymbol{u}^{(1)} + \left(\boldsymbol{x}, \boldsymbol{u}^{(2)}\right)\boldsymbol{u}^{(2)} + \cdots + \left(\boldsymbol{x}, \boldsymbol{u}^{(n)}\right)\boldsymbol{u}^{(n)}. \tag{H.12}$$

Accordingly, the vector $\boldsymbol{A}\boldsymbol{x}$ reads

$$\boldsymbol{A}\boldsymbol{x} = \lambda_1\left(\boldsymbol{x}, \boldsymbol{u}^{(1)}\right)\boldsymbol{u}^{(1)} + \lambda_2\left(\boldsymbol{x}, \boldsymbol{u}^{(2)}\right)\boldsymbol{u}^{(2)} + \cdots + \lambda_n\left(\boldsymbol{x}, \boldsymbol{u}^{(n)}\right)\boldsymbol{u}^{(n)}, \tag{H.13}$$

which is a **generalized Fourier series expansion** of $\boldsymbol{A}\boldsymbol{x}$ with respect to the basis $\{\boldsymbol{u}^{(i)}\}$. The result indicates that the matrix operation of $\boldsymbol{A}$ on $\boldsymbol{x}$ is equivalent to scholar multiplication of the eigenvalue λ_i to the ith term in the Fourier series expansion[2] of $\boldsymbol{x}$. Similar procedure can apply to the equation

$$\boldsymbol{A}\boldsymbol{x} = \boldsymbol{y}, \tag{H.14}$$

by which we obtain the solution of $\boldsymbol{x}$ as

$$\boldsymbol{x} = \frac{1}{\lambda_1}\left(\boldsymbol{y}, \boldsymbol{u}^{(1)}\right)\boldsymbol{u}^{(1)} + \frac{1}{\lambda_2}\left(\boldsymbol{y}, \boldsymbol{u}^{(2)}\right)\boldsymbol{u}^{(2)} + \cdots + \frac{1}{\lambda_n}\left(\boldsymbol{y}, \boldsymbol{u}^{(n)}\right)\boldsymbol{u}^{(n)}. \tag{H.15}$$

---

[1] A **unitary matrix** $\boldsymbol{U}$ is defined by that satisfying the condition of $\boldsymbol{U}\boldsymbol{U}^* = \boldsymbol{U}^*\boldsymbol{U} = \boldsymbol{I}$, where $\boldsymbol{I}$ is the identity matrix. The condition implies the distance-preserved property of $\boldsymbol{U}$, written by $\|\boldsymbol{U}\boldsymbol{u}\| = \|\boldsymbol{u}\|$ where $\boldsymbol{u}$ is an arbitrary vector in the space considered.

[2] A similar consequence is observed when a differential operator acts on a function that is written by Fourier series expansion. Acting a differential operator on the function is equivalent to multiplication of a specific wavenumber to each term of the Fourier series expansion.

Appendix I

Eigenspace Decomposition

I.1 Eigenspace of matrix

Let μ be an eigenvalue of a matrix $\boldsymbol{A}$ with the $n \times n$ size. In general, one or more eigenvectors belong to the eigenvalue μ. The set of these eigenvectors is denoted by $E(\mu)$, namely

$$E(\mu) = \{\boldsymbol{x} : \boldsymbol{A}\boldsymbol{x} = \mu \boldsymbol{x}\}. \tag{I.1}$$

If all the n eigenvalues of the matrix $\boldsymbol{A}$ take different values to each other, every set $E(\mu)$ has only one element. But this is not the case in general. For instance, the matrix

$$\boldsymbol{A} = \begin{bmatrix} 3 & 0 & 0 \\ 0 & 3 & 0 \\ 0 & 0 & 1 \end{bmatrix}, \tag{I.2}$$

has only two mutually different eigenvalues (i.e., 1 and 3), while

$$\boldsymbol{A} = \begin{bmatrix} 3 & 3 & 0 \\ 3 & 3 & 0 \\ 0 & 0 & 1 \end{bmatrix}, \tag{I.3}$$

has three eigenvalues (i.e., 0, 1, and 6). In the former case, two eigenvectors that belong to the single eigenvalue of $\mu = 3$ consists of the set $E(\mu = 3)$. It thus follows that the number of eigenvalues of $\boldsymbol{A}$ may be less than n, in which a set $E(\mu)$ is composed of more than one eigenvectors of $\boldsymbol{A}$. Now we are ready to enjoy the definition below.

Definition (Eigenspace):
Given a matrix $\boldsymbol{A}$, the set $E(\mu) = \{\boldsymbol{x} : \boldsymbol{A}\boldsymbol{x} = \mu \boldsymbol{x}\}$ is called an **eigenspace** that belongs to μ. ■

Example:
The 3×3 matrix $\boldsymbol{A}$ given by (I.2) has only two eigenvalues, $\mu = 1$ and $\mu = 3$. The corresponding eigenspaces read

$$E(3) = \left\{ \begin{bmatrix} a \\ b \\ 0 \end{bmatrix}, \begin{bmatrix} c \\ d \\ 0 \end{bmatrix} \right\}, \quad E(1) = \left\{ \begin{bmatrix} 0 \\ 0 \\ 1 \end{bmatrix} \right\}, \tag{I.4}$$

where the constants $abcd$ must satisfy all the three conditions of $\begin{bmatrix} a \\ b \end{bmatrix} \neq \begin{bmatrix} kc \\ kd \end{bmatrix}$ with k being an arbitrary real number, $\begin{bmatrix} a \\ b \end{bmatrix} \neq \mathbf{0}$ and $\begin{bmatrix} c \\ d \end{bmatrix} \neq \mathbf{0}$. ■

Note that when more than one elements are contained in an eigenspace $E(\mu)$, those elements are not uniquely determined. This fact has already been observed in the example abovementioned. The elements contained in $E(3)$ were represented by constants $abcd$ whose values are not specified. Any two vectors $\begin{bmatrix} a \\ b \\ 0 \end{bmatrix}, \begin{bmatrix} c \\ d \\ 0 \end{bmatrix}$ consists of the set $E(3)$ properly, provided that the two vectors are in linear independence as written by $\begin{bmatrix} a \\ b \end{bmatrix} \neq \begin{bmatrix} kc \\ kd \end{bmatrix}$ with arbitrary (but non-zero) k. Actually, infinitely many choices are available for the vector-representation of the elements of $E(3)$ such as

$$E(3) = \left\{ \begin{bmatrix} 1 \\ 0 \\ 0 \end{bmatrix}, \begin{bmatrix} 0 \\ 1 \\ 0 \end{bmatrix} \right\} \tag{I.5}$$

or

$$E(3) = \left\{ \begin{bmatrix} 1/\sqrt{2} \\ -1/\sqrt{2} \\ 0 \end{bmatrix}, \begin{bmatrix} 1/\sqrt{2} \\ 1/\sqrt{2} \\ 0 \end{bmatrix} \right\}, \tag{I.6}$$

and so on. Remind that in any choice, the number of elements is kept to be two. Importantly, the linear-independence condition imposed on the elements of an eigenspace $E(\mu)$ regulates the maximum number of elements that can be included in the eigenspace. Such the maximum number is referred to as the dimension of $E(\mu)$, designated by $\dim E(\mu)$. In the example above, we have

$$\dim E(3) = 2, \quad \dim E(1) = 1. \tag{I.7}$$

For general $n \times n$ matrices,

$$1 \leq \dim E(\mu) \leq n. \tag{I.8}$$

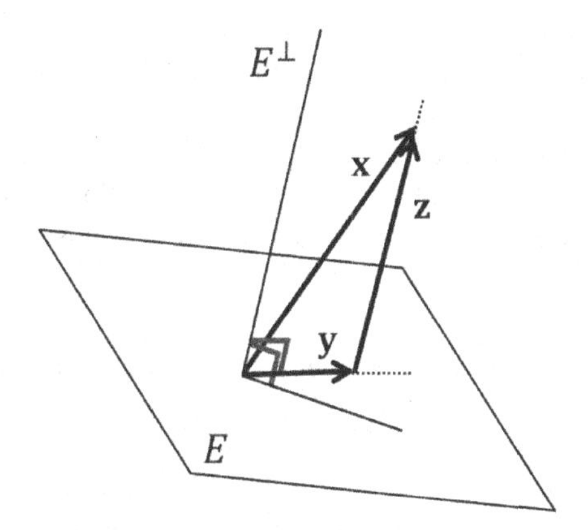

Figure I.1: Decomposition of $\boldsymbol{x}$ in the three-dimensional space, into the sum of $\boldsymbol{y} \in E$ (i.e., in the two-dimensional plane) and $\boldsymbol{z} \in E^{\perp}$ (i.e., in the straight line normal to the plane).

I.2 Direct sum decomposition

Meanwhile, we express the mutually different s eigenvalues of an $n \times n$ matrix $\boldsymbol{A}$ by $\mu_1, \mu_2, \ldots, \mu_s$, where $1 \leq s \leq n$. As well, eigenspaces that correspond to those eigenvalues are represented by $E(\mu_1), E(\mu_2), \ldots, E(\mu_s)$. The following theorem follows.

Theorem:
If the elements $\boldsymbol{x}_1 \in E(\mu_1), \boldsymbol{x}_2 \in E(\mu_2), \ldots, \boldsymbol{x}_s \in E(\mu_s)$ satisfy

$$\boldsymbol{x}_1 + \boldsymbol{x}_2 + \cdots + \boldsymbol{x}_s = \boldsymbol{0}, \tag{I.9}$$

all of them must be zero vectors written by

$$\boldsymbol{x}_1 = \boldsymbol{x}_2 = \cdots = \boldsymbol{x}_s = \boldsymbol{0}. \quad \blacksquare \tag{I.10}$$

This theorem suggests the fact that there exists no correlation between an element $\boldsymbol{x}$ of an eigenspace $E(\mu_i)$ and an element $\boldsymbol{y}$ of a different eigenspace $E(\mu_j)$ with $\mu_j \neq \mu_i$. For example, a vector $x \in E(\mu_1)$ cannot be represented as a linear combination of the elements that belong to $E(\mu_i)$ with $i \neq 1$. In terminology, we say that the eigenspace $E(\mu_1)$ of a matrix is **orthogonal** to other eigenspaces of the same matrix. This orthogonality holds for arbitrary pairs of the eigenspaces of a given matrix. As a consequence, an n-dimensional vector space V is written by the combination of mutually independent eigenspaces $E(\mu_1), E(\mu_2), \ldots, E(\mu_s)$ such as

$$V = E(\mu_1) \oplus E(\mu_2) \oplus \cdots \oplus E(\mu_s), \tag{I.11}$$

where the symbol $\oplus$ indicates to take a direct sum of two spaces. (I.11) is called the direct sum decomposition[1] of the space V with respect to the eigenspaces $E(\mu_i)$ of the matrix $\boldsymbol{A}$.

Another relevant concept to the direct sum decomposition is the notion of projection matrix. In order to give the definition of a projection matrix, let us suppose that an n-dimensional vector space V is a direct sum of two subspaces E and $E^\perp$. This assumption is written by

$$V = E \oplus E^\perp. \tag{I.12}$$

Under the assumption, an arbitrary element $\boldsymbol{x} \in V$ equals to a sum of a certain element in E, $\boldsymbol{y} \in E$, and that in $E^\perp$, $\boldsymbol{z} \in E^\perp$, as expressed by

$$\boldsymbol{x} = \boldsymbol{y} + \boldsymbol{z}, \quad \boldsymbol{y} \in E, \quad \boldsymbol{z} \in E^\perp. \tag{I.13}$$

Here, $\boldsymbol{x} \in V$ and $\boldsymbol{y} \in E$ are related to each other through the mapping defined below.

Definition (Projection matrix):
Let V be a vector space of $V = E \oplus E^\perp$ and consider an element $\boldsymbol{x} \in V$. The matrix $\boldsymbol{P}$ that yields $\boldsymbol{P}\boldsymbol{x} = \boldsymbol{y} \in E$ is called a **projection matrix** toward the subspace E of V. ■

Given a subspace E of an n-dimensional vector space V, we can find a set of orthonormal vectors of E, denoted by $e_1, e_2, \cdots, e_m$. Using the orthonormal set, a vector $\boldsymbol{P}\boldsymbol{x}$ that is obtained by acting $\boldsymbol{P}$ on $\mathbf{x} \in \mathrm{E}$ reads as

$$\boldsymbol{P}\boldsymbol{x} = (\boldsymbol{x}, e_1)\, e_1 + (\boldsymbol{x}, e_2)\, e_2 + \cdots + (\boldsymbol{x}, e_m)\, e_m. \tag{I.14}$$

Furthermore, we assume that the n-dimensional vector space V can be decomposed into a direct sum as

$$V = E_1 \oplus E_2 \oplus \cdots \oplus E_s. \tag{I.15}$$

Under the situation, interestingly, the identity matrix $\boldsymbol{I}$ is expressed by

$$\boldsymbol{I} = \boldsymbol{P}_1 + \boldsymbol{P}_2 + \cdots + \boldsymbol{P}_s, \tag{I.16}$$

where $\boldsymbol{P}_i$ is the projection matrix onto the space E_i. That (I.16) is true follows from the argument shown below. In terms of direct sum decomposition, an element $\boldsymbol{x} \in V$ is decomposed as

$$\boldsymbol{x} = \boldsymbol{x}_1 + \boldsymbol{x}_2 + \cdots + \boldsymbol{x}_s, \quad \boldsymbol{x}_i \in E_i. \tag{I.17}$$

Operating $\boldsymbol{P}_i$ on both sides, therefore, we obtain

$$\boldsymbol{P}_i \boldsymbol{x} = \boldsymbol{x}_i. \tag{I.18}$$

Substituting this result into (I.17), we obtain

$$\boldsymbol{x} = \boldsymbol{P}_1\boldsymbol{x} + \boldsymbol{P}_2\boldsymbol{x} + \cdots + \boldsymbol{P}_s\boldsymbol{x} = (\boldsymbol{P}_1 + \boldsymbol{P}_2 + \cdots + \boldsymbol{P}_s)\,\boldsymbol{x}, \tag{I.19}$$

which implies (I.16). It warrants noteworthy that (I.14) and (I.16) demonstrates the essential features of the projection matrix $\boldsymbol{P}$. These features give a clue to developing the theory of linear operators acting on infinite dimensional spaces.

[1]It is also called an orthogonal decomposition.

Index

Printed in the United States
by Baker & Taylor Publisher Services